MW00512686

BRASSEY'S WORLD MILITARY TECHNOLOGY

EXPLOSIVES, PROPELLANTS and PYROTECHNICS

BRASSEY'S WORLD MILITARY TECHNOLOGY

Series Editor: Colonel R G Lee, Royal Military College of Science, Shrivenham, UK

Incorporating Brassey's earlier *Land Warfare*, *Air Power* and *Sea Power* series, *Brassey's World Military Technology* encompasses all aspects of contemporary and future warfare. Faced with the complexity and sophistication of modern hardware, the fighting soldier needs to understand the technology to communicate his requirements to a design team. The *Brassey's World Military Technology* series provides concise and authoritative texts to enable the reader to understand the design implications of the technology of today and the future.

Titles currently available:

Aerospace Reconnaissance ISBN 1 85753 138 8
GJ Oxlee

Air Defence ISBN 0 08 034759 2
MB Elsam

Air Superiority Operations ISBN 0 08 035819 5
JR Walker

Amphibious Operations ISBN 0 08 034737 1
MHH Evans

Battlefield Command Systems ISBN 1 85753 289 9
MJ Ryan

Command and Control: Support Systems in the Gulf War ISBN 1 85753 015 2
MA Rice & AJ Sammes

Explosives, Propellants, Pyrotechnics ISBN 1 85753 255 4
A Bailey & SG Murray

Guided Weapons ISBN 1 85753 237 6
JF Rouse

Maritime Air Operations ISBN 0 08 040706 4
B Laite

Military Ballistics ISBN 1 85753 084 5
GM Moss, DW Leeming & CL Farrar

Military Rotorcraft ISBN 1 85753 325 9
P Thicknesse, A Jones, K Knowles, M Kellett, A Mowat and M Edwards

Noise in the Military Environment ISBN 0 08 035831 4
RF Powell & MR Forrest

Powering War ISBN 1 85753 0489
PD Foxton

Small Arms ISBN 1 85753 250 3
DF Allsop & MA Toomey

Strategic Offensive Air Operations ISBN 0 08 035805 5
M Knight

Submarines ISBN 0 08 040970 9
JB Hervey

Surveillance & Target Acquisition Systems ISBN 1 85753 137 X
MA Richardson, IC Luckraft, RS Picton, AL Rodgers & RF Powell

The Aerospace Revolution ISBN 1 85753 204 X
RA Mason

EXPLOSIVES, PROPELLANTS AND PYROTECHNICS

Professor A. Bailey *and* Dr S. G. Murray

Royal Military College of Science, Shrivenham, UK

BRASSEY'S
London

First English Edition 1989
Reprinted 1996
Reprinted 2000

UK editorial offices: Brassey's, 9 Blenheim Court, Brewery Road, London N7 9NT

A member of the Chrysalis Group plc

UK orders: Littlehampton Books, 10 – 14 Eldon Way, Lineside Estate, Littlehampton BN17 7HE

North American orders: Brassey's, Inc., 22841 Quicksilver Drive, Dulles, VA 20166, USA

Library of Congress Cataloging in Publication Data
Available

British Library Cataloguing in Publication Data
A catalogue record for this book is available from the British Library

ISBN 1 85753 255 4 Hardcover

Front cover: Close-up of the Sheridan Shillelagh Weapon System being fired. Courtesy of the Chrysalis Picture Library.

Printed in Great Britain by Redwood Books, Trowbridge, Wiltshire

Preface

This series of books is written for those who wish to improve their knowledge of military weapons and equipment. It is equally relevant to professional soldiers, those involved in developing and producing military weapons or indeed anyone interested in the art of modern warfare.

It is intended that the books should be of particular interest to officers in the Armed Services wishing to further their professional knowledge as well as anyone involved in research, development, production, testing and maintenance of defence equipment.

The authors of the series are all members of the staff of the Royal Military College of Science, Shrivenham, which is composed of a unique blend of academic and military experts. They are not only leaders in the technology of their subjects, but are aware of what the military practitioner needs to know. It is difficult to imagine any group of persons more fitted to write about the application of technology to the battlefield.

This Volume

In the hi-tech world of the battlefield much emphasis is directed towards ever increasing sophistication of weapons systems. However, it can be said that, almost without exception, a weapon system would be impotent without its explosive filling. It is true, also, that ever increasing importance is being placed on the safety and reliability of munitions.

This book introduces the reader to the science of technology associated with the three main classes of explosive: detonating high explosives, propellants and pyrotechnics. A basic introduction is given to both burning and detonating explosives; initiation is covered in some detail, covering both practical methods and hazards. Gun and rocket propellants are considered along with the other type of burning explosives, pyrotechnics. Military pyrotechnics are often high energy systems and the scope and nature of these materials are not generally known. Pyromechanisms are outlined in an attempt to clarify their position in the sphere of explosives. Finally, filling is considered because, technically, this is a difficult area in the production of the end product—battlefield munitions.

April 1989
Shrivenham

FRANK HARTLEY
GEOFFREY LEE

Contents

List of Tables

List of Illustrations

Introduction

Gunpowder was undoubtedly the first explosive, recorded in a Chinese manuscript, 1000 AD. It is of interest to note that this manuscript '*Wu Ching Tsung Yao*' was a treatise on military arts and that from that time, the science and technology of explosives has been driven forward by military requirements.

The next major event in the history of explosives came with the discovery of nitric and sulphuric acids. This allowed chemists to 'nitrate' the wealth of new compounds discovered by their science. Many nitrated materials were made and many chemists lost their lives in the process, as it was soon discovered that these materials were very energetic and often very unstable. However, in 1861 Alfred Nobel, the father of modern explosives, constructed a production plant for the manufacture of nitroglycerine at Heleneborg in Sweden.

In the same decade, nitrocellulose became available. This, together with nitroglycerine, gave the foundation for modern gun propellants, finally ousting gunpowder by the turn of the century. Gunpowder had already been replaced as a shell filling by picric acid (trinitrophenol) and then by TNT (trinitrotoluene) before the end of World War I.

After 1918, major research programmes were inaugurated to find new and more powerful explosive materials. From this programme came one of the most used military explosives of today, that is, cyclotrimethylenetrinitramine (RDX) also called Cyclonite or Hexogen. Since this time, other explosive materials of military use have been discovered and will be discussed within the relevant chapters. However, the vast majority of explosives and propellants used today consist of the materials mentioned above, RDX, TNT, nitroglycerine and nitrocellulose. This is something of a simplification since other explosive types used in far smaller quantities are vital to the operation of explosive devices. Thus, explosives are part of all munitions and obviously vital to their performance. They must be considered as a method of storing energy to be liberated on command to drive the weapon, be it a propellant driving the shell from the gun barrel or the high explosive shell filling driving the fragments formed from the shell case. It is the aim of this book to explain the science and technology of the explosives used today within the three main sub-groups of explosive materials: (high) explosives, propellants and pyrotechnics.

Finally, to complete the story of gunpowder. As related above, gunpowder has been replaced as a propellant and as a destructive filling however, it is far from obsolete. Since gunpowder is used in delays, ejection charges for carrier shells and igniter pads for propellants, it is still as vital to the operations of a modern army as it was in the Middle Ages.

1. The Chemistry and Physics of Explosives

The Nature of Explosions

Physical Explosions

An explosion is far easier to define than an explosive. Berthelot, in 1883, defined an explosion as *the sudden expansion of gases into a volume much greater than their initial one, accompanied by noise and violent mechanical effects.* Explosions are common in nature. At one end of the scale, a lightning strike may explosively split a tree trunk; at the other end, island volcanoes, such as Krakatoa, have exploded with such violence that waves have been created which swept over the oceans more than halfway round the world to cause immense damage by flooding when they impacted on the shore. In both cases, the explosions were due to the sudden vaporisation of water without any chemical reaction taking place. The bursting of a balloon or the rupture of a compressed gas cylinder remind us that heat is not a necessary prerequisite for an explosion nor is the presence of a recognised explosive substance. These physical explosions are nevertheless inherently destructive and capable of causing damage by air blast and by the propulsion of fragments at high velocity.

Nuclear Explosions

Nuclear explosions are a second class of explosions. These result in the sudden release of enormous quantities of heat by fission or fusion processes. The heat causes the rapid expansion of the air surrounding the device and also vaporises material in the close vicinity which adds to the effect. The radioactive elements are not themselves *explosives* in the conventional sense but the device is triggered by the use of conventional explosives.

Chemical Explosions

The third class of explosion is called a chemical explosion. This is caused by the extremely rapid reaction of a chemical system to produce gas and heat. The reaction is usually a combustion process and is often accompanied by the production of smoke and flame.

Explosives

There are many substances known to chemists which are so unstable that they may explode with little or no stimulus and are often used for party tricks. If we simply define an explosive as a substance capable of causing an explosion, we shall include many substances which have no practical value and constitute little more than a nuisance in the laboratory. The legal definition in Great Britain is given in the Explosives Act of 1875, where an explosive is '*a substance used or manufactured with a view to produce a practical effect by explosion*' such as gunpowder, nitroglycerine and TNT. Practical explosives must be inert to substances with which they may come into contact, including air and moisture, and they must be thoroughly stable under the expected conditions of storage and use. At the same time, they must be sufficiently sensitive to be initiated by convenient means. For a practical explosive, the minimum energy for initiation should be small in comparison with the total amount of energy released on explosion. These various properties collectively form the two essential requirements for any useful explosive, namely, SAFETY and RELIABILITY. Many candidate explosive substances are ruled out for practical use by the application of these criteria.

To be useful, as implied by the Explosives Act, an explosive must also be EFFECTIVE, that is it must be able to do work on its surroundings. This takes place by the expansion of the hot high pressure gas produced in the chemical reaction against the surroundings resulting in airblast, propulsion of fragments, lift and heave effects. An explosive may therefore be defined as a substance or mixture of substances which when suitably initiated, decomposes explosively with the evolution of heat and gas.

High explosives are a very convenient source of power. When one kilogramme of a detonating explosive is initiated in a bore hole, it will generate power at a rate of 5000 megawatts over the duration of the reaction. Such a charge occupies less than one litre of volume, and may be stored for years. Yet at the touch of a button it may release its energy when and where required.

The History of Explosives

Black Powder

Explosives, in the usual sense of that word, do not occur in nature. Although explosions are not unknown in the natural world, they are usually physical phenomena caused by the sudden vaporisation of water in some way. The earliest man-made explosive is properly referred to as black powder, but is more commonly called gunpowder because of the main use to which it has been put during most of its existence. Black Powder is an intimate mixture of saltpetre (potassium nitrate) with charcoal and sulphur. The last two have always been available in many parts of the world, sulphur occurs naturally in volcanic regions and charcoal is easily made by heating wood. Saltpetre does not occur naturally in anything like a pure state, but it forms an efflorescence on the ground when animal or vegetable matter decomposes *in situ*. This process is a long one and occurs only in certain climatic conditions, which are experienced in China and the Middle East among other places. At some unknown time and place, one or more questing minds found that charcoal and sulphur, which are only moderately flammable when mixed with saltpetre, which is non-flammable, give rise to a mixture which is not only flammable but violently so. On the balance of the slim documentary evidence, the discovery was made by the Chinese, perhaps around AD 700, and there are indications that by AD 1000 they had developed weapons of war utilising black powder, though not what we now call guns.

News of the discovery spread only slowly round the world, but eventually it reached the ears of scholars in Europe, one of whom was the man whose name is most closely associated with black powder, the Englishman Roger Bacon (born circa 1214, died 1292). Bacon graduated c.1240 and became a Franciscan friar c.1257, but most of his life was devoted to languages, mathematics, optics and experimental science. Amongst these activities he experimented in the purification of saltpetre and made some effective black powder: this much is known because he included the details, partly in cypher, in a scientific work which he wrote for the attention of the Pope. The experiment does not figure prominently in Bacon's biography, which is based largely on his own writing, in the virtual absence of cross-references. He seems deliberately to have avoided spreading the knowledge further, for reasons which can only be guessed at.

However, certain of Bacon's learned contemporaries, such as Albertus Magnus (c.1200–1280) also knew about black powder and were less reticent about it. Hence, before the end of the 13th century, it was being used in war for the breaching of fortifications. By 1320 it was being used in guns, with enormous consequences for the future of the world. The composition and manufacture of powder were slowly improved over the next few centuries. Its first recorded use for civil engineering was in the dredging of the bed of the River Niemen in Northern Europe, 1548–72. Mineral blasting is said to have been pioneered by Gaspar Weindl in Hungary in 1627 and to have been introduced in England in 1689 by Thomas Esply senior, in the Cornish tin and copper mines. Demolitions by black powder feature in British history in the murder of Lord Darnley at Kirk O'Field in 1567, an abortive attempt on the Houses of Parliament in 1605 and the destruction of fortifications during and after the Civil War of 1642–6. The building of the Languedoc Tunnel in France, 1666–81 involved what was probably the first major use of black powder for public works tunnelling.

Nobel's Nitroglycerine

The ultimate limitations of black powder as a blasting explosive became frustratingly apparent during the great growth of communications in Europe and North America in the mid-nineteenth century, especially in such formidable enterprises as the building of the trans-Alpine railway tunnels. Fortunately, a remedy was forthcoming in the discovery, in 1846, of the liquid explosive nitroglycerine (NG) by the Italian Professor Ascanio Sobrero (1812–88). Sobrero realised that NG was far more powerful than black powder but he soon lost interest in it, and it was taken up a few years later by the Swedish inventor Immanuel Nobel. By 1863, the latter, together with his third son Alfred Bernhard Nobel (1833–96), had set up a small factory for NG manufacture at Heleneborg near Stockholm. Attempts by the Nobel family to market the product were hampered, however, by the serious disadvantages of NG: its inconsistency of response to initiation by ordinary flame which is the standard method for black powder, and the difficulty and danger of handling the explosive in its normal liquid state. This latter gave rise to numerous accidental explosions, one of which destroyed the Nobels' factory in 1864. Alfred Nobel, a very industrious science graduate, eventually solved these problems by the invention of the fulminate detonator (1864) and the discovery of an effective absorbent (Kieselguhr) (1867). The twin products, 'guhr dynamite' and the detonator, provided an invaluable combination far superior to black powder for most of the needs of civil engineers and miners, and quickly amassed a fortune for Alfred Nobel. He did not stop there, however: in 1869 he combined NG with mixtures of nitrates and combustibles to form a new class of explosives known at that time as 'straight dynamites'. In 1875 he made a further definitive discovery, namely that nitrocellulose mixed with NG forms a gel, and from this stemmed the invention of blasting gelatine, gelatine dynamites and later, in 1888, the first 'double-base' gun propellant, ballistite.

Nitrocellulose

Nitrocellulose (NC) had been first prepared at about the same time as NG, but being a more complex material it was developed in stages by different workers, notably T.J. Pelouze in 1838 and C.F. Schonbein in 1845. Schonbein used NC for mineral blasting and experimentally in guns, demonstrating that as a propellant it was potentially better than gunpowder, because it gives virtually smokeless combustion. Factory production of NC began at Faversham, Kent, in 1846, but in the following year an explosion destroyed the plant and two other NC disasters occurred in France. In Austria attempts to manufacture NC in a safer, more suitable form were persevered with, and in 1863 production was resumed in England at Stowmarket, Suffolk, but an explosion there and another in Austria in 1865 halted the work in both countries. During these years Sir Frederick Abel (1827–1902) had been working on the instability problem for the British Government at Woolwich and Waltham Abbey, and in 1865 he published his solution, which consisted of pulping, boiling and more thoroughly washing the product. This breakthrough was followed by another in 1868 when

Abel's assistant E.A. Brown discovered that dry compressed guncotton (highly nitrated NC) could be detonated by the use of a fulminate detonator. This paved the way for the effective use of NC as a military and commercial high explosive.

Propellant Development

Because of the problems with NC, Schonbein's early use of it as a gun propellant was not successfully followed up for many years. In 1864 the Prussian Schultze developed a propellant from nitrated wood, which, though unsuitable for rifles, proved useful for shotguns. In 1870 the Austrian manufacturer of this powder patented its partial gelatinisation with ether and ethanol. During the next two decades the manufacture of 'smokeless' shotgun propellants was steadily improved, but it was not until 1884 that the Frenchman Paul Vieille produced the first NC propellant safe for use in rifled guns. This was named Poudre B. Eventually, all such powders, based on NC alone, became known as single-based types to distinguish them from Nobel's ballistite. In 1889, a year after the ballistite patent, a rival product of similar composition to ballistite was patented by the British Government in the names of Abel and Dewar, and was called Cordite. In its various modifications it was to remain the preponderant British service propellant until the 1930s when triple-base propellants, containing nitroguanidine, were developed for ordnance. Cordite remained in use for British small arms ammunition until after World War II, when a change to single-base types was made.

Development of Commercial Explosives

Returning to commercial blasting explosives, the vast range of NG-based explosives continued to dominate the market throughout the latter part of the 19th century and the first half of the 20th, despite attempts to substitute other bases such as chlorates, perchlorates or liquid oxygen. The period 1880–1914 was marked by determined attempts, particularly in France and Britain, to eliminate unsuitable explosives from coalmines. This was the result of long-term trends. In 1800 British coal production was 10 million tons annually. Frequent underground explosions of natural methane or suspended coaldust killed hundreds of miners until checked by the invention of the Davy Lamp in 1815. Production rose rapidly, reaching 50 million tons in 1850 with the increased use of black powder, and 133 million tons by 1875 with the additional help of dynamite. However, the growing use of explosives brought a corresponding increase in the number of gas and dust explosions, with appalling casualty totals. Scientists gradually deduced the physical mechanisms by which explosives can ignite methane, and 'testing galleries' similar to large boilers were built and successfully improved. Those explosives which, even after modification, failed the gallery tests were prohibited by law from use in coal-mines. They included dynamite and black powder. They were replaced by 'permitted explosives', which tended increasingly to be based on ammonium nitrate, and other limitations were also imposed, notably on the weight fired in each shot. By 1913 British coal production reached an all-time peak of 287 million tons, consuming more than 5000 tons of explosives annually, and by 1917 92 per cent of these were ammonium nitrate-based. After the Great War the use of explosives declined with total coal production, and also due to machine cutting, which by 1938 accounted for 59 per cent of production compared with 8 per cent in 1913. Today the manufacture of permitted explosives in Britain is only a small, though stable percentage of the total, reflecting both the fall of underground coal production to 110 million tons by 1981 and the maximum use of coal-cutting machines.

The Development of Military Explosives

The Great War of 1914/18 saw enormous production of explosives for munitions. High explosive shells had been commonly filled with picric acid since Turpin's patents of 1885 and the British adoption of it in 1888 under the name Lyddite. Around 1902 both the Germans and the British had experimented with TNT, first prepared by Wilbrand in 1863. On the outbreak of war the British hastily followed the Germans in making the change to TNT, but almost immediately found that demand for it outstripped supply, so they changed to a mixture of TNT and ammonium nitrate called amatol. During the war some 238,000 tons of TNT were produced in Britain alone. The production of tetryl for exploders in shells also began in 1914 and accelerated greatly.

After the war, work on new explosives continued in many countries, and much interest was centred on the compounds known in Britain as RDX and PETN. RDX had been first prepared by the German Henning in 1899, but its immense potential had been overlooked until 1920, when it was patented by Herz. Efforts were made to overcome the difficulty and high cost of manufacture, and problems of sensitiveness, in order to obtain explosive mixtures far superior to those used in the Great War. The Germans concentrated more on PETN. During World War II TNT and amatol were again the main fillings for British shells and bombs, but RDX, produced at the Royal Ordnance Factory at Bridgwater from 1939, was added to these compositions insofar as supplies permitted. There was a greater usage of aluminium as an additive (to give a high blast effect) than had been possible in World War I. The generic composition 'torpex' (TNT/RDX/aluminium,) became common. During the war, the marginally superior analogue of RDX named HMX was first produced but not in great quantity.

Since 1945 it has become recognised in military circles that no practical explosive superior in performance to RDX and HMX is ever likely to be produced at an acceptable cost. British munition fillings have been largely standardised on the RDX/TNT 60/40 mixture. Research has mainly been aimed at optimising compositions for special applications, and identifying and solving safety problems. The use of RDX and HMX has been extended into the field of high-performance propellants for both guns and rockets. Incorporation of RDX, HMX or PETN into oily or polymer matrices has produced plastic or flexible explosives for demolitions, while other polymers provide tough, rigid, heat-resistant compositions for conventional missile warheads and for the implosion devices used in nuclear weapons.

The Development of Commercial Explosives

Developments in the field of commercial blasting since the 1939–45 war have been more radical than those in the military one. The manufacture of NG, a delicate and dangerous process upon which much of the explosives industry has depended since Nobel's time, had been revolutionised in 1929 by the Schmidt continuous nitration method, and in the 1950s it was further improved by the introduction of the injector nitration method combined with continuous centrifugal separation. However, the cost of NG-based blasting explosives remained comparatively high, so in the early 1950s attention turned to finding a more cost-effective alternative. The answer was provided, as it had been to other problems, by ammonium nitrate, this time in admixture with fuel oil, and the acronym ANFO was bestowed on the product. In America annual consumption of ANFO increased from 268,000 tons in 1960 to 759,000 tons in 1970. A second revolution occurred a few years after ANFO with the introduction of slurry explosives. These are also based on ammonium nitrate used in the form of a saturated aqueous solution gelled by a cross-linking agent, and with excess ammonium nitrate held in suspension. Various fuels are added, such as carbonaceous material or aluminium, and there must be some further additive, e.g. TNT, methylamine nitrate or fine aluminium powder, to render the mixture sensitive to initiation. The idea of aqueous slurry explosives dates back to the turn of the century but its practical form is credited to Melvin A Cook and dated 1957. The great increase in the use of ANFO and slurries, combined with the decreased use of explosives in underground coal mining, has had a depressing effect on the manufacture of NG-based blasting explosives, but these still have a vital place in the demolition industry for breaking metal and masonry, and for various other applications where precision is more important than large scale and severe economy.

In recent decades, the commercial uses of explosives have extended beyond the crude destruction methods of older times. Rapid changes in society and industry, for example, have necessitated the demolition of strong, fairly modern buildings such as factories, power stations and blocks of flats, often in close proximity to vulnerable facilities. This has forced the development of new and critical blasting techniques. The growth of the off-shore oil industry has required the penetration of the seabed, and cutting of submerged metal structures. Such requirements have led to the adaptation of hollow charge techniques, developed originally for military purposes, and the use of modern explosive compositions formerly considered too expensive for non-military users. Some engineering projects now include explosive bolts, and explosives are also used for forming shapes in metal and for welding metal plates. This proliferation of applications has justifiably led to the concept that explosives technology is now a branch of engineering in its own right.

Accidental Explosions

Early Problems

As with most of the dangerous commodities used by mankind, explosives have caused innumerable deaths and injuries over the years, but with each new category of accident the application of human intelligence sooner or later finds a way to avoid a repetition. This is, of course, subject to avoidance of human error, and perhaps the best advice available to all who handle explosives is the old British Army adage; 'We must never forget that the ultimate purpose of an explosive is to explode'.

During the centuries when technology advanced only slowly, the use of black powder for blasting was particularly dangerous because there was no reliable way to convey the initiating flame to the charge with a delay time sufficient for the operator to retreat safely. Trains of powder along floors, or quills or rushes filled with powder, often caused a premature explosion and a casualty, an eventuality immortalised by Shakespeare in the phrase 'hoist with his own petard'. In the Cornish mines of the 19th century, where miners were particularly reckless, such occurrences were commonplace. Then in 1831, William Bickford of Tuckingmill, Camborne, invented the 'miner's safety fuse', which abruptly changed the situation worldwide and remains in use today in a virtually unchanged form. As we have seen, the mass explosions of NG in the 1860s impelled Alfred Nobel to find a remedy in the form of dynamite, and the accidents with NC in the same period led to Abel's work which elevated that explosive to its rightful position of great utility.

19th-century Reforms

In the latter part of the 19th century the increasing production of explosives led to many forms of accident. In Britain, particular concern was caused by the explosion of two powder magazines at Erith, Kent, which killed 13 people, and there followed one at Messrs. Ludlow's in Birmingham which killed 53. The Government, in the reforming tradition of the time, assumed the responsibility for preventing recurrences if possible, and appointed Colonel Sir V.D. Majendie to investigate the Birmingham incident. Afterwards he guided the resulting legislation which became the Explosives Act of 1875. Under the Act, Colonel Majendie was appointed the first permanent Inspector of Explosives to administer its provisions. He had power to inspect all magazines and factories and to see that operations were carried out in a reasonably safe manner. The result of the Act was a dramatic cut in the average number of persons killed *per annum* in British explosives factories;

1868–70	43
1871–74	32
1878–87	7.5

Notable Accidents

Other Governments followed suit and the international problem was brought under control. It was found, however, that the carrying of explosives on board ships posed higher risks than those on land, because of the concentration of the cargo, the ease with which fire spreads and the general proximity of port facilities to densely populated areas. A ship explosion at Santander, Spain, in 1893 killed hundreds of people, and on 6 December, 1917, during the Great War, a ship carrying 2500 tons of explosives blew up off shore at Halifax, Nova Scotia, Canada, killing 1963 people and destroying the town centre.

In Britain the war brought a number of accidental explosions, the largest being at a TNT works in Silvertown, East London. The highest death toll, however, was 134 killed at an explosives factory at Chilwell, Nottinghamshire, on 1 July, 1918.

A series of explosions at a United States naval ammunition depot in New Jersey in 1936, was caused by lightning and involved 1500 tons of explosives and 21 deaths. It was significant in that it led to the adoption of the quantity-distance (Q–D) concept, and rules were then applied in the United States limiting the quantity of explosive which could be stored within a given distance of other explosives or of inhabited buildings. This concept is widely applied today but not always without its problems, especially where explosives ships, port facilities and population centres exist together.

A new cause of disaster was discovered in 1921. Some 4000 tons of ammonium nitrate fertiliser, stored in a silo at a chemical works at Oppau, near Mannheim, Germany, exploded without warning at 7.33 a.m. on 21 September. The village of Oppau was largely destroyed and out of its 5000 population 561 were killed and 1500 injured. The effects of the explosion were felt up to 20 miles away.

Wartime Accidents

World War II, like its predecessor, brought a huge increase in the amount of explosives existing around the world, often handled or stored under difficult conditions by inexperienced personnel. In Britain there were inevitable accidental explosions in factories, ammunition dumps and on trains, but their frequency was minimised by experience gained in the previous war, and often by cases of personal bravery. An illustration of the latter, and also of unforeseen cause and effect, occurred in Cambridgeshire on the night of 1 June, 1944. A freight train loaded with 400 tons of aircraft bombs was approaching Soham when it was found that the timber bed of the leading wagon was on fire. The bombs which it was carrying contained five tons of TNT. The crew stopped the train, coolly unhitched the remaining trucks, then restarted the locomotive and drew the burning wagon forward as far as Soham Station. There its load exploded, but due to the marginal gap which had been opened up the remainder of the train load did not. Two railwaymen were killed, and the station and 15 houses were destroyed. The truck, as well as the locomotive tender, was demolished, but subsequent investigation showed that it had previously been used for carrying sulphur, and it was surmised that a spark from the engine had ignited the residue, thus starting the fire. Appropriate recommendations were made, as always to avoid a repetition.

The largest explosion in the country's history occurred accidentally at RAF Faulds, an underground bomb store near the village of Hanbury in an agricultural part of Staffordshire. On 27 November, 1944 some 3500 tons of bombs detonated almost simultaneously, killing 68 people and destroying not only the depot but parts of Hanbury Village half a mile away. The cause is believed to have been the use of a wrong tool on a tetryl exploder. The crater, 100 feet deep and several acres in extent, is a tourist attraction to this day. The same year of 1944, marking as it did the climax of the War, was a bad one for explosives accidents. On 14 April a fire and explosion on a ship in Bombay docks killed more than 1000 persons, and on 17 July at Port Chicago, California, 1500 tons of explosives, mainly torpex, in railcars and a ship's hold, blew up killing 320 and injuring 390.

Post-war Problems

Another bad year occurred in 1947. This time the cause was not military explosives but, as at Oppau, large quantities of ammonium nitrate fertiliser. In France, 21 people died and 100 were injured at Brest when 3000 tons of fertiliser exploded, and on 16/17 April of the same year, when two ships caught fire at Texas City, USA, some 2280 tons of nitrate which they were carrying exploded, causing 552 fatalities and leaving more than 3000 injured. It is said that these disasters served, ironically, to draw anew the attention of the explosives industry to the capacity of ammonium nitrate to detonate *en masse*, and hence paved the way for the introduction of ANFO and slurries a few years later.

In recent decades there have been no incidents world wide of accidental explosions causing casualities on the scale of the 1940s, and in general the industries responsible for the manufacture, storage, transportation and use of explosives can be said to have put their house in order. Indeed, the chemical engineering industry has attracted more opprobium with disasters involving other dangerous substances (e.g. Flixborough 1974, Movosibirsk 1979, Bhopal 1984) than has been caused by conventional explosives. The discovery that ammonium nitrate 'cooks off' when exposed to a conflagration had echoes in Britain in January 1977 in connection with another oxidising salt, sodium chlorate, which until then had not been thought capable of mass explosion. A warehouse at the Braehead container depot, Renfrew, Scotland, containing 1700 drums of the substance accidentally caught fire and, ten minutes later, exploded, propelling debris up to five miles, slightly injuring 13 people and causing £6 million damage. As a result of this, the storage of sodium chlorate is now subject to much stricter controls. Incidents will always continue to happen with conventional explosives and other dangerous materials, but experience, wisdom, care and new technology can keep fatalities to a minimum.

The Nature of Explosives

Basic Features

Explosive systems exist in a wide range of physical forms. They may be single compounds such as TNT or RDX or they may be mixtures of substances. These mixtures may consist of chemical compounds or simple elements and they may be liquids, solids, gases or, as frequently happens with commercial explosive systems, two-phase systems. Solid or liquid explosives are often called CONDENSED explosives.

The most usual chemical reaction which occurs in the explosion process is combustion or oxidation. Here, the fuel elements, usually carbon and/or hydrogen, react with the oxidising elements, normally oxygen or a halogen such as chlorine or fluorine. Condensed explosives contain the necessary oxidiser to permit reaction to take place in the absence of air and are thus 'self contained'. In contrast, ordinary fuels such as paraffin, wood or coal are not capable of rapid combustion unless they are finely divided and provided with copious quantities of air, or more specifically, oxygen. The corollary of this is that because explosives contain their own oxygen as well as fuel, when combustion takes place, the total amount of energy they release is considerably less than that from an equal weight of a straight forward fuel. For example, 1 kilogramme of TNT releases 4080 kilojoules of energy whereas the combustion of 1 kilogramme of petrol in an adequate supply of air will release more than 30,000 kilojoules. The usefulness of an explosive derives more from the speed at which it releases its energy than from the quantity of energy released.

Explosive Mixtures

As we have already seen, explosives exist in a number of forms and the simplest way to classify them is into mixtures and single chemical compounds.

The simplest mixture we can consider is that of 2 volumes of hydrogen gas and 1 volume of oxygen gas. Complete mixing occurs naturally, and when initiated with a spark a violent explosion ensues with the production of water

$$2H_2 + O_2 \longrightarrow 2H_2O$$

This reaction results in a decrease in the number of gas molecules and at first appears to be contrary to the definition of an explosive which is that a large volume of gas should be produced. However, no less than 13,260 joules of heat per gramme of mixture are released by the combustion of hydrogen. The expansion of the product water vapour due to the heat much more than makes up for the reduction in the number of gas molecules on combustion and an explosion results. Though gaseous fuel/air explosives have been proposed for mine clearance, they are not practical military explosives and 'condensed' mixtures are more commonly encountered.

An example of a condensed explosive mixture is black powder or gunpowder. Roger Bacon (1214–92) gave its composition as:

Potassium nitrate	37.5	per cent
Charcoal	31.25	per cent
Sulphur	31.25	per cent

This mixture was slow burning and not very powerful. Developments in production methods took place which brought the constituents into much more intimate contact and hence speeded up the combustion rate. 'Corned' (granulated) black powder burns much more rapidly than the older material. Another improvement came when it was realised that there was not enough oxygen held in the potassium nitrate oxidiser to combust all the fuel elements. The formula for a modern black powder is typically

- ▷ Potassium nitrate 75 per cent
- ▷ Charcoal 15 per cent
- ▷ Sulphur 10 per cent

The sulphur, as well as being a fuel element, reduces the ignition temperature of the mixture. Black powder was invented long before the advent of modern chemistry and even now it is not possible to describe its combustion adequately in a simple chemical equation. An over-simplified approximation to the reaction is as follows:

$$4KNO_3 + 7C + S \longrightarrow 3CO_2 \text{ (gas)} + 3CO \text{ (gas)} + 2N_2 \text{ (gas)}$$
$$+ K_2CO_3 \text{ (solid)} + K_2S \text{ (solid)}$$

Thus only about 43 per cent of the explosive is converted into gaseous products which is about a quarter of that produced by an equal quantity of an efficient modern explosive. Moreover, the solid products retain a considerable proportion of the heat produced on explosion, further reducing the efficiency of the black powder.

A much more efficient and commonly encountered condensed explosive mixture is used commercially as a 'prilled' powder or as a slurry and about 25,000 tons was produced in the United Kingdom alone in 1986. The reaction may be represented by the equation:

$$25NH_4NO_3 + C_8H_{18} \longrightarrow 8CO_2 + 59H_2O + 25N_2$$

The products of this explosive mixture are entirely gaseous giving a very cheap and efficient product.

Single Explosive Compounds

The development of the organic chemical industry in the mid-19th century was stimulated by the search for synthetic dyestuffs which had a greater fastness and colour range than those extracted from natural materials. Soon a vast range of compounds was available. Many of the coloured compounds resulted from the treatment of precursor materials with mixtures of nitric and sulphuric acids. This treatment produced compounds containing one or more ($-NO_2$) groups (nitrogroups) many of which were explosives in their own right. The oxygen for combustion is held in the ($-NO_2$) group.

Such groups may be attached to oxygen atoms, carbon atoms or nitrogen atoms in the molecules and are called 'nitrate', 'Nitro' or 'nitramine' compounds respectively. Examples of explosives of each type are shown below:

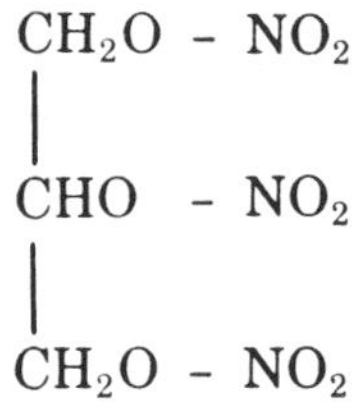

$CH_2O - NO_2$
$CHO - NO_2$
$CH_2O - NO_2$

Glycerol trinitrate

(nitroglycerine)

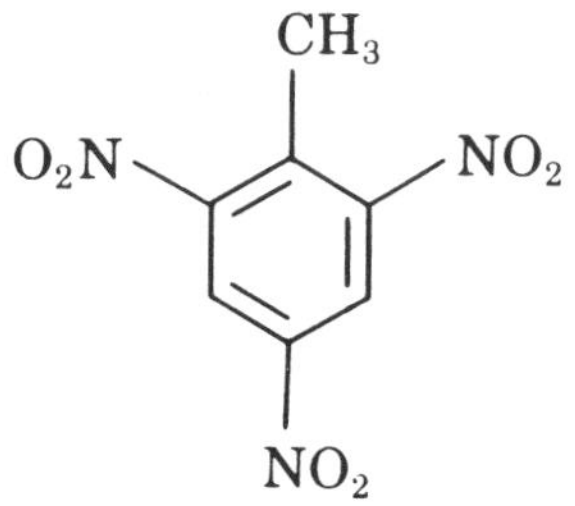

CH_3, O_2N, NO_2, NO_2

Trinitrotoluene

(TNT)

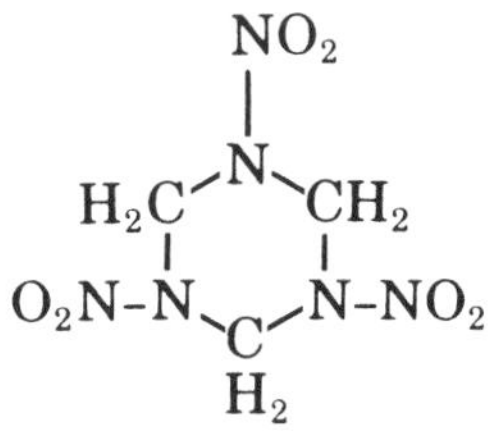

NO_2, H_2C, N, CH_2, O_2N-N, C, $N-NO_2$, H_2

Cyclotrimethylene trinitramine

(RDX)

These molecules will explode violently if stimulated in an appropriate manner. The molecular structure breaks down on explosion leaving, momentarily, a disorganised mass of atoms. These immediately recombine to give predominantly gaseous products and evolve a considerable amount of heat called the Heat of Explosion.

▷ $C_3H_5N_3O_9 \longrightarrow 3CO_2 + 2.5H_2O + 1.5N_2 + .25O_2$
Nitroglycerine

▷ $C_7H_5N_3O_6 \longrightarrow 3.5CO + 2.5H_2O + 1.5N_2 + 3.5C$ (solid)
TNT

▷ $C_3H_6N_6O_6 \longrightarrow 3CO + 3H_2O + 3N_2$
RDX

Performance of Explosives

The effectiveness of an explosive depends on two factors. The first is the *amount* of energy available in the explosive and secondly, the *rate* of release of the available energy when the explosion occurs. To measure the effectiveness of different explosives a variety of performance parameters may be used such as:

- ▷ Heat of explosion
- ▷ Temperature of explosion
- ▷ Pressure of explosion
- ▷ Power index
- ▷ Rate of burning
- ▷ Detonation velocity
- ▷ Detonation pressure.

These parameters may be measured experimentally or calculated from theory. Theoretical calculations are more convenient and are useful in comparing the relative performance of one explosive with another but in an explosion the precise conditions are not known so it is doubtful whether absolute performance figures for an explosive can be calculated precisely. Where such absolute values are required then they must be determined by experiment.

Heat of Explosion (Q)

When an explosive is initiated either to rapid burning or to detonation, its energy is released in the form of heat. The heat so released under adiabatic conditions is called the Heat of Explosion and determines the work capacity of the explosive. For high explosives and for propellants in the breach of a gun, the heat is measured or calculated under constant volume conditions (Q_v). For rocket propellants, where the products of combustion may expand freely into the atmosphere, it is more appropriate to use conditions of constant pressure and the heat obtained is Q_p.

The heat of explosion is simply the difference between the heat of formation of the products of explosion and the heat of formation of the explosive compound itself. The heats of formation of gases such as carbon dioxide, carbon monoxide and water are well known and have often been measured. The heats of formation of chemical compounds may be calculated from a knowledge of the individual bond energies between the atoms which make up the explosive molecule thus enabling the heat of explosion to be calculated easily. Table 1 lists values of Q for various common explosives:

TABLE 1.1 Calculated Heats of Explosion for Some Explosives (Water as Gas)

	joules per gramme
Nitroglycol (EDGN)	6730
Nitroglycerine	6275
PETN	5940
RDX	5130
HMX	5130
Tetryl	4350
TACOT	4015
TNT	4080
DATB	3805
Lead azide	1610
RDX/TNT 60:40	4500

The experimental determination of heats of explosion has yielded rather variable results, depending on the experimental conditions and the oxygen content of the particular explosive. For instance, when the mixture of atoms formed during an explosion recombines to form the gaseous products, the precise mix of CO_2, CO, H_2O etc. depends on the loading density in the test vessel and this factor also affects the measured heat of explosion.

Oxygen Balance

From the formulae of the three explosives given earlier, we can see immediately that nitroglycerine has more than enough oxygen in its structure to oxidise fully the fuel elements of the molecule. RDX is somewhat oxygen deficient since the carbon has been oxidised to carbon monoxide only. TNT is very oxygen deficient; not only has its carbon only been oxidised as far as carbon monoxide but some carbon has not been oxidised at all. This is the reason why TNT gives a black sooty smoke when it is detonated.

Unlike explosive mixtures, where the oxygen content can be adjusted by varying the proportions of the constituents until the fuel and oxidiser elements are balanced, simple chemical compounds are rarely perfectly balanced. The oxygen balance (Ω) of an explosive is defined as:

> 'the percentage by weight of oxygen, positive or negative, remaining after explosion, assuming that all the carbon and hydrogen atoms in the explosive are converted into CO_2 and H_2O.'

All the explosive classes above — nitrate esters, nitrocompounds and nitramines — contain only the elements carbon hydrogen oxygen and nitrogen and are called CHNO explosives having the general formula:

$$C_a H_b N_c O_d$$

Simple algebra shows that the oxygen balance is given by the formula:

$$\Omega = (d - 2a - b/2) \times \frac{1600}{M}$$

where M is the relative molecular mass of the explosive. Using the formula we find the oxygen balance shown below:

	Oxygen balance
Nitroglycerine	+3.5 per cent
RDX	-22 per cent
TNT	-74 per cent

The Effect of Oxygen Balance on Heat of Explosion

When there is exactly enough oxygen in the explosive to fully oxidise the carbon and hydrogen to carbon dioxide and water (i.e. $\Omega = 0$) then the heat of explosion will be optimal. Any deviation from perfect oxygen balance either positive or negative, will lead to a lower heat of explosion. Figure 1.1 shows the dependence of the heat of explosion on oxygen balance. EGDN, for which $\Omega = 0$ has the highest heat; nitroglycerine, with its positive oxygen balance ($\Omega = +3.5$) has a smaller heat as do all the other explosives with negative oxygen balances.

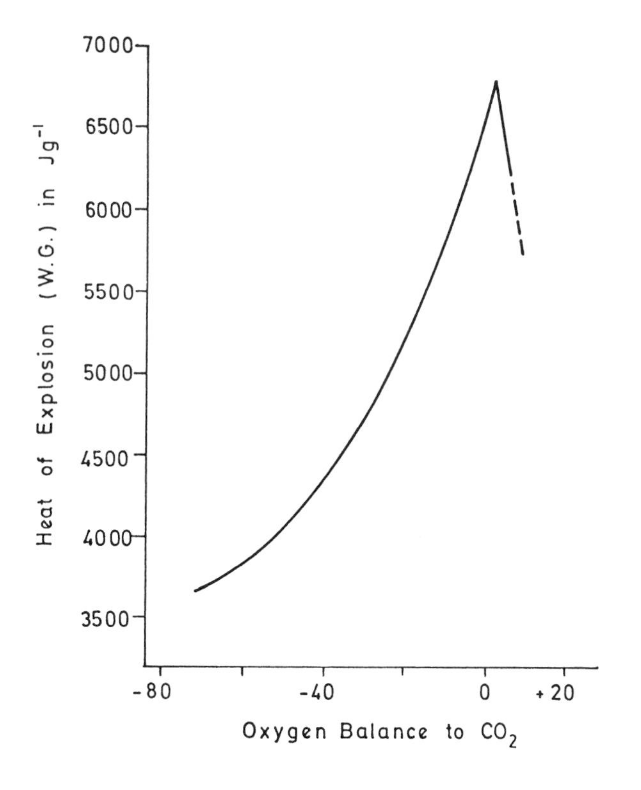

FIG. 1.1 Effect of oxygen balance on the heat of explosion

TNT is heavily oxygen deficient. Its performance can be improved by mixing it with ammonium nitrate which is oxygen rich to form the 'amatols'. An ammonium nitrate/TNT 80:20 mixture has $\Omega = 0$. In this case, though a high Q value is obtained, the detonation velocity suffers, being only 5080 m s^{-1} for amatol of density 1.45 g cm^{-3} compared with 6950 m s^{-1} for TNT at 1.58 g cm^{-3}.

In blasting gelatine, the oxygen deficiency of the 8 per cent cellulose nitrate content is compensated for by the addition of 92 per cent of nitroglycerine. Hence the heat of explosion is higher than that of either component alone and in addition, the mixture forms a gel of high density and helps to account for the high velocity of detonation which is 7900 m s^{-1} at 1.55 g cm^{-3}.

Composition of the Gaseous Products of Explosion

To calculate the heat of explosion and other explosive performance parameters a knowledge of the composition of the gaseous products of explosion is essential. This may be calculated from the equilibrium constants of the water-gas and other reactions but it is a tedious process. Recently, data banks and software have been compiled which allow computer calculations to be made.

A good approximation to the experimentally determined composition can easily be obtained however, by applying a set of rules developed by G. B. Kistiakowsky and E. B. Wilson during World War II. The rules can be applied to any CHNO explosive provided it has an oxygen balance of greater than (–40).

The rules are:

1. All carbon atoms in the molecule are oxidised to carbon monoxide.
2. Any unused oxygen atoms then oxidise the hydrogen atoms to water.
3. Any remaining oxygen atoms are used by oxidising their stoichiometric amount of carbon monoxide to carbon dioxide.
4. Any nitrogen atoms go to nitrogen gas (N_2).

The rules have been applied to the following examples:

EXAMPLE 1

What are the gaseous products of RDX($C_3H_6O_6N_6$) which are formed on detonation? (Ω (RDX) = –22 per cent)

From the above rules:

1. $3C \longrightarrow 3CO$ (3 oxygen atoms left)
2. $6H \longrightarrow 3H_2O$ (all oxygen used up)
3. No oxygen left
4. $6N \longrightarrow 3N_2$

The overall reaction is therefore:

$$C_3H_6O_6N_6 \longrightarrow 3CO + 3H_2O + 3N_2$$

For explosives with a greater oxygen deficiency than (–10) the above rules were modified as follows:

1. All hydrogen atoms in the molecule are oxidised to water.
2. Any unused oxygen atoms then oxidise carbon to carbon monoxide.
3. All nitrogen atoms go to nitrogen gas.

EXAMPLE 2

What are the gaseous products of TNT($C_7H_5O_6N_3$) which are formed on detonation (Ω(TNT) = –74%)

From the above rules we get:

1. 5H $\longrightarrow$ $2.5H_2O$ (3.5 oxygen atoms left)
2. 3.5C $\longrightarrow$ 3.5CO (all oxygen used up)
   5H 3.5 carbon atoms are left unoxidised
3. 3N $\longrightarrow$ $1.5N_2$

The overall reaction is therefore:

$$C_7H_5O_6N_3 \longrightarrow 2.5H_2O + 3.5CO + 3.5C\text{ (solid)} + 1.5N_2.$$

Temperature of Explosion

The maximum temperature which the gaseous products of explosion can reach if no heat is lost to the surroundings is called the Temperature of Explosion (T_e) and is often used when calculating the ability of an explosive or propellant to do work. T_e is found to lie between 2500 and 5000°C for military high explosives. It can be calculated quite easily if the quantities and nature of the gaseous products and the heat of explosion (Q) are known from the equation:

$$T_e - T_a = \frac{Q}{\Sigma \bar{c}}$$

T_a is the ambient temperature — generally taken to be 25 °C and $\Sigma \bar{c}$ is the sum of the mean molar heat capacities of the explosion products over the temperature range T_a to T_e. These are readily available from standard thermochemistry tables.

In practice, the heat capacities of molecules are higher, the more atoms there are in the gas molecules. It follows therefore, that if the product molecules formed on explosion can be kept small (e.g. produce CO rather than CO_2) then higher values of T_e may be achieved for the same heat of explosion (Q). This is achieved by adjusting the oxygen balance of the explosive so that it is somewhat negative. This will of course reduce the heat of explosion so that in practice a compromise is established. For gun propellants it is important not to have too high a temperature of explosion or else erosion of the gun barrel will reach unacceptable levels.

Gas Volume

The volume of the gaseous products of explosion (V) is generally calculated at a pressure of 1 bar and a temperature of 0°C (273K) and generally lies between 700 and 1000 cm^3 for military explosives. Inspection of the equation for combustion gives the number of moles of gas produced and since the volume of a mole of any gas occupies the same volume under standard conditions (22,400 cm^3) the value of V can be calculated readily.

EXAMPLE

Calculate the volume, V, for the detonation of RDX, given that the equation for reaction is:

$$C_3H_6O_6N_6 \text{ (RDX)} \longrightarrow 3CO + 3H_2O + 3N_2$$

$$1 \text{ mole of RDX} \longrightarrow 9 \text{ moles of gas}$$

$$222 \text{ g RDX} \longrightarrow 9 \times 22{,}400 \text{ cm}^3 \text{ of gas}$$

$$1 \text{ g RDX} \quad \text{gives} \quad \frac{9 \times 22{,}400 \text{ cm}^3 \text{ of gas}}{222}$$

$$V = 908 \text{ cm}^3 \text{ g}^{-1}$$

Pressure of Explosion

The pressure of explosion (P_e) is the maximum static pressure which may be achieved when a given weight of explosive is burned in a closed vessel of fixed volume. The pressures attained are so high that the Ideal Gas Laws are not sufficiently accurate and have to be modified by using a co-volume (α):

At high pressures

$$P_e (V^* - \alpha) = nRT_e$$

where V^* is the volume of the closed test vessel, n is the number of moles of gas produced per gramme of explosive and R is the universal gas constant.

EXAMPLE

Calculate the pressure of explosion of P_e for the burning of 10 g of RDX in a bomb of volume 10 cm^3, given that n = 0.045 moles per gramme of RDX, T_e = 4255K and α = 0.54.

$$P_e = \frac{0.0405 \times 8.31 \times 4255}{(1 - 0.54)}$$

$$= 3113 \text{ MN m}^{-2}$$

$$= 31 \text{ kbar approx.}$$

Pressures of explosion are of a much lower order of magnitude than detonation pressures.

2. The Explosion Process: Detonation Shock Effects

Burning and Detonation

Intrinsic Energy

In Chapter 1 we saw that practical explosives are mixtures or single compounds and that all but a few of them contain oxygen in some form, together with the fuel elements carbon and hydrogen. We saw also that nitrogen is usually present as a chemical companion to the oxygen. The system is thus capable of producing, almost instantaneously, a large quantity of the gases carbon dioxide, carbon monoxide, water and nitrogen. Because such oxidation reactions are exothermic, the system also produces much energy, which is treated as being initially in the form of heat and is called the heat of explosion, designated Q. Since the explosive material is a self-contained chemical system, and because energy cannot be created, it is clear that the energy is contained within the explosive from the time of its manufacture, waiting to be released by some initiatory stimulus.

Two Combustion Processes

When an explosive charge is initiated, the energy is released by one of two possible combustion processes, burning or detonation. Most explosives are capable of either process, depending on the method of initiation and the conditions under which it occurs, particularly the degree of confinement. In practice these factors are regulated in a way which ensures that the explosive behaves in the desired manner. Because of this, it is sometimes suggested that explosives can be divided into two categories, detonating and deflagrating or rapidly burning, but this is a misleading over-simplification, the more so as certain very useful explosive compounds are used increasingly in both roles. It is better to divide all explosive processes into either burning or detonation, and to bear in mind that in situations where the conditions cannot be regulated, i.e. in unforeseen accidents, the question as to which process happened is often decided by pure chance rather than the type of explosive involved. There is an element of unpredictability in the behaviour of explosives under all but tried and proven conditions.

Burning

Virtually all explosives burn vigorously when ignited in a dry, unconfined state, an exception to this are water-based slurries. Burning can also occur in a confined state, since explosives do not rely on an external supply of oxygen for their combustion. Burning comprises a series of chemical reactions which take place in a zone or zones at or just above the surface of the explosive. When the surface of the material is thus being converted into gases, the surface itself can be regarded as receding in consecutive layers. This concept is expressed in Piobert's Law (1839): The surface of a burning propellant recedes layer by layer in a direction normal to the surface.

The size of the discrete piece of explosive does not affect the concept: it can be tiny grain of small arms propellant, or the propellant grain of a solid rocket motor weighing tens of kilogrammes. The temperature of each layer is brought in turn to the ignition point by means of heat radiated or conducted from the reaction zone(s) into the solid material. Some heat is also evolved by the relatively slow decomposition of the material just before it reaches its ignition point, the model assumes that the surface is non-porous, which does not apply to certain propellant compositions. The rate at which the surface recedes depends on the rate of heat transfer into the material. In turn, this depends on the temperature at the burning surface, the thermal conductivity of the material, its transparency to radiation and inversely on its thermal stability.

Rate of Regression

The rate of regression, or 'linear' burning rate, is designated r. For a given explosive the main factor affecting r is the pressure P obtaining at the surface at a given instant, since the effect of pressure is to thin the reaction zone and thereby accelerate the flow of heat into the explosive. In 1862 de Saint Robert deduced that the relationship, for black powder, between r and P was given by

$$r = \beta P^{2/3}$$

where β is called the coefficient of the burning rate and depends on the units of r and P. Later, Paul Vieille found that the smokeless powders and other new explosives of the Nobel era gave different values for the index and he generalised the expression as

$$r = \beta P^{\alpha}$$

which is known as Vieille's Law (1893).

Burning Rate Index

The index α, known as the burning rate index, has to be determined experimentally using an apparatus called a Strand Burner. Strands of propellant are burned at various pressures and the measured values of r are plotted against P. Values of α for different propellants and other explosives vary from about 0.3 to more than 1.0. If the index is exactly 1.0, the plot is a straight line. An index higher than 1.0 gives a steepening curve, less than 1.0 gives a flattening one. The index is of great importance in the internal ballistics of guns and rockets. It also indicates the tendency or otherwise of an explosive to detonate in certain circumstances, as will be seen later. The value of α for conventional gun propellants is about 0.8–0.9, while that for certain rocket propellants is as low as 0.3.

Because of the pressure dependence of the linear burning rate, the effect of heavy confinement on a burning explosive is very great. A strand of typical gun propellant, ignited at one end in the open, burns quietly at around 5 mm s^{-1}, but when confined in the breech of a gun, where the pressure can rise some 4000 times higher, the burning rate is about 400 mm s^{-1}. Thus a solid grain 2 mm thick would be consumed in 2.5 ms at maximum pressure. For the purpose of internal ballistics this burning rate does not need to be exceeded, but for any application requiring a more extreme burning rate the appropriate technique is to combine heavy confinement with the provision of porosity in the grains, so that the flame enters the particles and consumes them faster than would surface burning. By such means an ultimate linear burning rate of about 500 m s^{-1} is achievable.

Mass Rate of Burning

A different, but related, way of expressing the rate of burning is the mass rate of burning, designated m. This is the mass of explosive consumed in unit time. Consider a grain of explosive, of any shape, burning all over its surface. If the surface area is A and the linear burning rate is r, the volume of explosive consumed in unit time is $r \times A$. Hence the mass consumed in unit time is $r \times A \times \rho$ where ρ is the density. This would apply to, say, a single large propellant grain in a rocket motor or equally to an aggregate of smaller grains in the breech of a gun. Now, it can be shown by geometry that a large number of small grains has a much greater surface area per unit mass, (specific surface), than a single grain of the same weight. In a gun, therefore, A is large. The charge of grains is ignited quickly over its entire surface, so the rapid generation of gas causes the pressure P to rise. The linear burning rate r follows suit, and r and P then rise interdependently to high values within one or two milliseconds. The net result is the more or less explosive effect known as deflagration.

A Surface Phenomenon

The burning of a solid explosive, therefore, is essentially a surface phenomenon, as is the case with most other combustible solids. The main difference is that explosives do not need a supply of air to sustain their burning. Confinement of burning paper, for example, will extinguish the flame, but confinement of an explosive does the opposite, it speeds up the reaction and may lead to an explosion.

Detonation

Nature of a Detonation

As stated earlier, burning is one of the two alternative combustion processes by which explosives release their energy. The other is detonation. The first notable feature of detonation is a shock wave which passes through the explosive material without being much affected by the relative position of the surface. The second outstanding feature of detonation is its great speed compared with any burning process. The velocity of the shock wave in solid or liquid explosives is between 1800 and 9000 m s^{-1}, an order of magnitude higher than that of a fast deflagration and two or three orders higher than an average one. Whereas, in burning, the rate at which the material decomposes is governed by the rate of heat transfer into the surface, the rate of decomposition of the material in the wake of a detonation wave is limited only by the velocity at which the material can transmit the wave.

Basic Methods of Initiating a Detonation

The practical initiation of explosives takes place under conditions regulated to produce either burning or detonation as required. The burning of most explosives is quite easily started by simple ignition, and once started is difficult to stop. Detonation is less easy to achieve in most cases. Certain conditions have to be provided so that the shock wave is formed and will propagate through the explosive material without attenuating and fading. Although the velocity and pressure of the wave are enormous once it is established, it is sometimes delicate and unpredictable in its early stage.

Initiation to detonation can occur in one of two ways:

- ▷ Burning to detonation.
- ▷ Shock to detonation.

Burning to Detonation

Burning to detonation, as the name implies, consists of a transition from the burning process already described: an ignition of the explosive followed by an abrupt acceleration of the flame front until it becomes transformed into a shock wave. The conditions under which this can happen are various, and also the readiness with which different explosives display the phenomenon varies greatly. Perhaps the easiest case to visualise is when an explosive is confined in a tube and ignited at one end. The gas generated cannot escape easily, so pressure tends to build up at the burning surface. If the explosive has a burning rate index of more than 1.0, or if, as is sometimes the case, the index is higher at high pressure than at low pressure, pulses will be generated, and these may accelerate the linear burning rate to sonic velocity. A shock wave will then be formed and the transition to detonation is complete. A zone of total chemical decomposition and formation of gases follows the wave until the charge is completely consumed.

Burning to detonation can be brought about not only by confinement in a tube but also by the presence of a critical mass of explosive. Thus one stick of dynamite may merely flare when ignited whereas a bundle of similar sticks may well burn to detonation. A loose heap of explosive can provide self-confinement and thus either burn or detonate, according to its size and to a property called explosiveness which will be discussed in a later chapter.

Shock to Detonation

The alternative way of initiating detonation, called shock to detonation, requires the action of a shock wave from an already detonating charge, called the donor, on the charge in question, the receptor charge. The two charges usually need to be in contact, or nearly so. When the shock wave passes through the receptor charge, the explosive material undergoes compression and adiabatic heating. This liberates some of its chemical energy, which has the effect of changing the natural deceleration of the wave into acceleration. Hence the pressure in the wave front is increased, more energy is released from the explosive and the wave continues to accelerate until it reaches the characteristic velocity of sound in the shock-compressed medium of the explosive. A burst of light (as revealed by high speed cine photography) signals the onset of detonation. However, if the shock wave from the donor charge is too weak, or if other conditions are unfavourable, the entering wave will fail to accelerate and will die out, leaving the bulk of the receptor charge chemically unchanged.

Although the final result of the two initiation systems is identical, there is a wide difference of time scale between them. In a designed burning-to-detonation system, such as a blasting detonator or a delay fuse, the delay may be of milliseconds. In an accident situation, for example if a pile of explosive is being disposed of by uncontrolled burning, it may convert to detonation quite randomly, seconds or even minutes after ignition. Compared with either of these two cases, the shock-wave may have to run a distance of millimetres or centimetres into the receptor charge before reaching detonation velocity, but this causes a delay of only a few microseconds, so there is a difference of three orders of magnitude at least. Given the proper conditions the shock-to-detonation method is the surest and most convenient way in which to initiate a main charge to detonation. However, from a wider viewpoint, shock-to-detonation is not an independent alternative to burning-to-detonation, because it requires a donor charge, and that charge can only be initiated by a burning-to-detonation system at the outset.

Functions and Classification of Explosives and Pyrotechnics

In practice, explosives are employed to fill certain roles. The nature of the role determines whether burning or detonation is required. The explosive will then be caused to function under conditions regulated to ensure that it behaves in the desired manner. Explosives can therefore be classified under the role in which they are normally employed. Explosives which are normally caused to detonate are called high explosives (HEs). Explosives which normally function in a burning mode are sometimes called low explosives. However, this term is now discouraged and because such compositions are mostly propellants of various kinds, they are nowadays called propellant explosives or simply propellants.

We can include in our scheme of functions the class of materials known as pyrotechnics, although they do not share all the characteristics of explosives which are set out in this Chapter. Pyrotechnics are mixtures of oxidising and reducing solids, capable of self-sustained combustion at rates which differ greatly from one composition to another. They are designed to produce special effects which supplement or simulate those produced by conventional explosives, and are outlined in part 3 of Table 2.1.

TABLE 2.1 Functions of Explosives

1. ***High explosives*** detonate to:
   Create shock waves
   Burst
   Shatter
   Penetrate
   Lift and heave
   Create airblast
   Create underwater bubble pulses

2. ***Propellants*** burn to:
   Propel projectiles and rockets
   Start I.C. engines and pressurise other piston devices
   Rotate turbines and gyroscopes

3. ***Pyrotechnics*** burn to:
   Ignite propellants
   Produce delays
   Produce heat, smoke, light and/or noise

While some explosive compounds can be assigned exclusively to the category of high explosives, others are more versatile. Nitrocellulose is present in many commercial high explosives, but is also the universal ingredient of all conventional gun propellants. Nitroglycerine is also present in many commercial high explosives and in many propellants, while RDX, hitherto regarded simply as a high explosive, is also being incorporated in more and more propellant compositions.

Another way of classifying explosives is by the readiness with which they are ignited and exploded, that is, their sensitiveness to initiation. (The difference in meaning between the words sensitivity and sensitiveness is a fine one and in this instance is irrelevant). Substances which are readily ignited or detonated by a small mechanical or electrical stimulus are called primary explosives: those which are not readily initiated thus, and therefore require the influence of an impinging shock wave to initiate them, are called secondary explosives. Propellants are not normally initiated by either of these methods in practice, but by the application of flame. Combining the classifications, we can describe both the input normally applied to an explosive and the response expected of it:

- ▷ A primary high explosive can be detonated easily.
- ▷ A secondary high explosive can be detonated, but less easily.
- ▷ A propellant explosive is not required to detonate at all.
- ▷ Whilst it is probably true that all primary explosives are capable of detonation, some are not required to do so in use, but only to deflagrate.

It must be stressed that the above statements are requirements, not firm predictions as to the behaviour of explosive materials under all conditions. For instance, in accident situations or under enemy attack, propellants may detonate, and secondary explosives, given the wrong kind of initiatory stimulus, may burn instead of detonating. The requirements presuppose that the explosive system will function as it is designed to.

Partition of Energy

Propellants

Consideration of explosion processes reveals that the energy released by an explosive undergoes several conversions and partitions within the brief time-span of the explosion. For instance, in a gun the energy of the deflagrating propellant is first released as internal energy of the gases, manifested as high temperature and pressure. As soon as the projectile begins to move up the bore, work is being done on it, and it acquires kinetic energy at the expense of the internal energy of the gases. Further internal energy is lost as the gases themselves accelerate and acquire kinetic energy. When the projectile clears the muzzle, the gases vent and collide with the atmosphere, transferring their kinetic energy to it in the form of a minor blast wave. The remaining energy of the gases is lost directly to the atmosphere and to the material of the gun in the form of sensible heat. In general little of the energy of a deflagration is converted into wave energy.

Detonations in Air

Taking now the case of a high explosive detonating in air, an appreciable proportion of the released energy is contained in the shock wave which passes through the charge. This wave spreads laterally, and at the far end of the charge frontally, into the surrounding air, but decelerates abruptly and is overtaken within a few charge diameters by the expanding gas front. The gas front compresses the atmosphere ahead of it and creates a powerful blast wave, containing much of the original energy of the explosive. Work has been done by setting the air molecules in motion, but it is accomplished more by the internal energy of the expanding gases than by the energy of the detonation shock wave.

Confined Detonations

If the same explosive, instead of detonating in air, is inserted in a hole drilled in solid rock and detonated there, the situation is somewhat different, because the gas is not free to expand straightaway. The detonation shock wave is transmitted more efficiently by a dense solid such as rock. It travels freely through the material, but the impulse which it imparts is not of a type which, by itself, can propel heavy pieces of rock over a distance and thus do measurable work. Instead, it produces intense compression for a very short time, and this tends to cause plastic and elastic flow in a homogeneous hard material. Hence the energy of the wave is largely dissipated as frictional heat. The usefulness of the wave lies in the fact that a brittle material will fail under intense compression, and even a tougher one will fail in tension when the initial compression phase of the wave is suddenly reversed by reflection. These extreme forces produce spalling and cracking effects in even the strongest materials, and the gases are then able to expand and heave the broken mass in any required direction. Thus the energy released by a detonation is partitioned between the shock wave and the internal energy, which is the work capacity of the gases which expand behind it. For some effects, such as breaking isolated boulders, only the shock wave energy is of use, while for others, like blowing a crater in loose ground, it is the heaving action which is important. Each separate application requires a characteristic partition of the available energy. At a given detonation velocity, every high explosive displays a specific partition of its energy, and it is therefore possible to choose the explosive best suited to a particular task.

Measuring the Partition of Energy

The best way to measure the partition is to detonate a charge under water: the detonation wave moves rapidly away and is well clear before the gas pressure fully expands the resulting bubble at a lower order of velocity. Measurements can thus be made of the separate energies of the detonation wave and the bubble. Such experiments show that the maximum proportion of the energy which goes into the shock wave is no more than about 50 per cent, and that the explosives which display this proportion are those with a high velocity of detonation.

Detonation Velocity and Pressure

Because of the partition of energy requirement, the velocity (D) at which the detonation shock wave proceeds through a charge is an important parameter of the explosive material. It can be predicted by calculation and measured experimentally, and although a maximum value can be assigned to a particular explosive, depending ultimately on its thermochemical properties, practical results may be considerably lower, for reasons outlined below.

▷ *Effect of density of loading* (Δ)

Since a detonation wave proceeds through the body of the explosive, the energy which it releases within and behind itself will depend on the mass of explosive traversed per unit area of the wave front. Hence the more mass that is concentrated into a given volume of explosive, the more energy the wavefront can release in order to sustain itself at a high velocity. Provided the charge is of reasonable diameter and well confined (see below), the velocity of detonation appears to be almost exactly proportional to the loading density and is covered by the relationship:

$$D_1 = D_2 + 3500\ (\Delta_1 - \Delta_2)$$

where D_1 = velocity of detonation at density Δ_1
and D_2 = velocity of detonation at density Δ_2

This formula applies to explosive compounds or mixtures of them, and the result is that plots of D against Δ for different explosives form more or less parallel straight lines. The achievement of high loading density has importance in munitions in addition to its effect on the velocity. High density is necessary if the maximum amount of explosive is to be contained in a compact munition. In commercial blasting, high density loading of shot-holes is usually an advantage, but not always so: it depends on the burden to be blasted.

To calculate the approximate velocity of detonation for a given explosive at a particular loading density, Marshall's formula is used:

$$D \text{ (in metres per second)} = 430\sqrt{nT_d} + 3500\ (\Delta - 1)$$

where n is the number of moles of gaseous products per gramme of explosive detonated,

T_d is the temperature of detonation, expressed in kelvins

and Δ is the charge density in g cm^{-3}.

This formula gives good agreement with observed results although the calculation of T_d from thermochemical data is necessarily only approximate. It should be noted that both of the foregoing formulae apply only to explosive compounds or mixtures of such compounds. They do not apply to explosive mixtures of fuel and oxidant: the detonation velocity of these falls off above a critical density, and if compressed too much they become undetonable. This is known as dead-pressing, or desensitisation.

▷ *Effect of diameter of charge*

The velocity of detonation falls below its maximum when the diameter of the charge is below a certain value, if the degree of confinement is also small. This diameter varies for different explosives, according to their sensitivity. For sensitive compounds like NG and PETN it is a few millimetres, for less sensitive ones like TNT, ANFO and slurries it is between 10 and 30 centimetres.

▷ *Effect of confinement*

In general, the stronger the confinement the higher is the detonation velocity. The effect of confinement becomes more marked as the diameter of the charge is reduced.

▷ *Effect of strength of detonator*

Explosive charges and initiating systems used in military munitions are designed to ensure that the fully velocity of detonation is achieved in every round, for maximum effect. However, the more *ad hoc* combinations of detonator and main charge used in commercial blasting operations may result in the detonator imparting an inadequate shock wave to the receptor charge, and the velocity of detonation will then be markedly decreased. Nitroglycerine-based explosives, commonly used for rock blasting, but not in military detonating munitions, are capable of two very different velocities. The use of too weak a detonator may reduce the achieved velocity from more than 6000 m s^{-1} to a low as 2000 m s^{-1}.

The corollary of all the variabilities of detonation velocity is that when quoting velocities it is necessary to state at least the loading density to which a particular figure applies. Where any other doubt is implicit, the diameter, confinement and strength of detonator should also be stated. For military explosives, the velocities quoted are usually the maximum allowed by all the variables. In any case, such compositions, being of high performance, are less affected by the variables. The velocities of commercial explosives are usually quoted, in this country, for an unconfined column of explosive 1¼ inches (32 millimetre) in diameter and initiated by a No. 6 detonator. These conditions often produce a velocity far below the maximum achievable by other means, as shown in the previous paragraph. Table 2.2 shows velocities for commercial explosives and Table 2.3 shows values for some military explosives. Table 2.2 is assembled from various data and shows velocities which probably approach the maximum achievable. Sales literature produced by manufacturers tends to quote a range of velocity for any given explosive, to cover all causes of variation.

The velocity of detonation shock waves is equal to the velocity of sound (*C*) within the explosive material at the temperature and pressure existing in the shock front, plus the velocity of the material (*w*) as it moves forward in the wave. This fact is known as the Chapman-Jouguet condition:

$$D = C + w$$

where *C* is about one-third higher than the speed of sound in the charge before it is compressed by the detonation wave. For a typical high explosive, say TNT, *c* is about 5400 m s^{-1} in the conditions of the wave and *w* is about 1500 m s^{-1}, so *D* is about 6900 m s^{-1}.

The Experimental Measurement of Detonation Velocity

The experimental measurement of detonation velocities has been greatly improved in recent years with the introduction of electrical methods and high-precision timing. A detonation shock front has a high electrical conductivity compared with the undetonated explosive, so circuits can be set up in the explosive which feed high-voltage pulses into a counter-timer. For routine quality control in explosives factories, however, a much older technique is used, known as the Dautriche method. This is a comparative method, relying on a known velocity for detonating fuze. This velocity, of course, has to be checked when necessary by an absolute method such as above. The following is an outline of the Dautriche method.

The explosive under test is contained at the desired density of loading in a steel tube and fired by the detonator, see Figure 2.1. Two detonators are inserted at points *A* and *B* in the tube which are at a measured distance *L* metres apart, and a length of detonating fuze of known velocity of detonation connects the two detonators together. The centre portion of the fuze is laid on a plastic plate *P*, the exact centre point of the fuze *E* being marked on the plate. As the detonation wave of the explosive travels along the tube it fires the two detonators in turn, causing waves of detonation travelling in opposite directions to pass around the fuze. Where these two waves meet at F an indentation occurs on the plastic plate and the distance EF is then measured.

Then if D_2 is the velocity of the detonation of the fuze, the velocity of detonation of the explosive is given by:

$$D_1 = \frac{D_2 L}{2EF}$$

This formula may be easily derived thus: the time taken for the explosion wave to travel from A to F in the fuze must be equal to the time taken for the explosion wave to travel from A to B in the main explosive and then B to F in the fuze. Hence:

$$\frac{AE}{D_2} = \frac{L}{D_1} + \frac{BF}{D_2}$$

or

$$\frac{AE + EF}{D_2} = \frac{L}{D_1} + \frac{BE - EF}{D_2}$$

But $AE = BE$, hence:

$$D_1 (BE + EF) = D_2 L + D_1 (BE - EF)$$

$$\therefore \quad D_1 = \frac{D_2 L}{2EF}$$

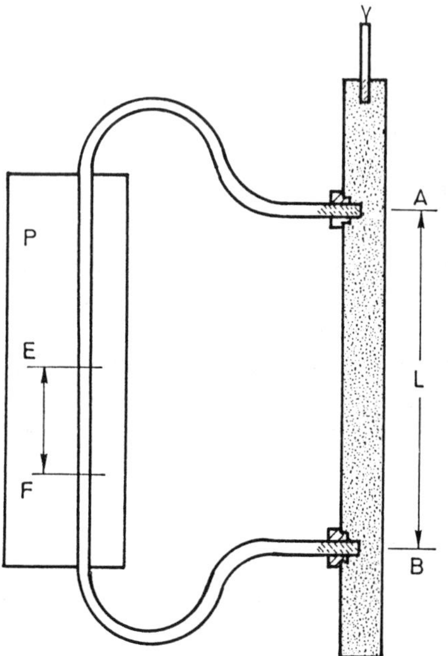

FIG. 2.1 Diagram of Dautriche Apparatus

TABLE 2.2 DETONATION VELOCITIES OF SOME COMMERCIAL EXPLOSIVES

Type	*Density* $g\ cm^{-3}$	*Detonation velocity* $m\ s^{-1}$	*Remarks*
Blasting gelatine (92% NG)	1.55	7900	Initiated with No.8 detonator
Polar ammon gelatine dynamite (50% NG)	1.5	6850	
Gelignite (26% NG)	1.45	6400	
Plaster gelatine	1.5	6300	
TNT powder (18% TNT)	1.1	4850	
Water-based slurries	1.4-1.7	3-4500	Depends on composition and diameter
ANFO	0.8	2-3000	Depends on diameter
Permitted powder explosive (10% NG, 12% sodium chloride)	0.7	1800-2000	

Detonation Pressure

The peak dynamic pressure in the shock front is called the detonation pressure (ρ) of the explosive. An empirical method of calculating it is due to Cook, as follows:

$$\rho\ (\text{kbar}) = \Delta D^2 \times 2.50 \times 10^{-6}$$

where Δ is the charge density in g cm^{-3}

and D is the velocity of detonation in m s^{-1}.

EXAMPLE

Given the velocity of detonation of HMX at 1.50 g cm^{-3} as 7400 m s^{-1}, calculate its detonation pressure (ρ) at the same density.

Then

$$\rho = 1.50 \times 7400^2 \times 2.5 \times 10^{-6}\ \text{kbar}$$
$$= 205\ \text{kbar}$$

The calculated detonation pressures of some high explosives are given in Table 2.3. It should be noted that it is traditional in explosives technology to use the kilobar as a unit of detonation pressure, regardless of the nature of the medium. Strictly, the bar and kilobar should only be used in connection with fluids, so some modern literature on shock waves quotes pressures in gigapascals (GPa), where 1 GPa is equal to 10 kbar. In the present work we are adhering to tradition.

Fragmentation Munitions

Detonation pressure is the main factor in determining the property of explosives called brisance, or shattering effect. It is derived from the French

TABLE 2.3 Detonation Parameters for some Military Explosives

Type	$D/m\ s^{-1}$	*at* $\Delta/g\ cm^{-3}$	$\rho/kilobar$ *calc.*
Secondary explosives			
HMX	9110	1.89	392
RDX	8440	1.70	300
RDX/TNT 60/40	7900	1.72	268
DATB	7520	1.79	253
Nitroglycol	8100	1.50	246
Nitroglycerine	7700	1.60	237
Tetryl	7160	1.50	192
TNT	6950	1.57	190
Primary explosives			
Lead azide	4500	3.8	192
Mercury fulminate	4500	3.3	167
Lead styphnate	4900	2.6	157

word *briser* = to break. The property of brisance enables a high explosive to break its container, such as a shell or bomb, into small fragments which fly at high velocities. There is no close analogy between this process and the ejection of a projectile from a gun. The latter depends on the relatively steady production of hot gas by the propellant, whereas fragmentation relies on the shock wave to break up the case and begin the acceleration, subsequently aided by the expanding gas. In modern munitions the ways in which the fragments are produced are various. Whereas a HE shell or mortar bomb is a plain cylinder which has to be strong enough to withstand the shock of discharge, a missile warhead or an anti-personnel grenade may have a liner of preformed fragments, or its case may be grooved on the inside to produce a similar effect. Fragmentation is a complex topic and the following is only a rough guide.

The fragment size, whether randomly produced or preformed, is optimised for the type of target intended. An anti-personnel grenade projects very small fragments, sometimes pieces of notched wire liner: they are required only to incapacitate an enemy, not kill him, and should not be capable of wounding the thrower at only a little greater distance. Fragments from warheads, shells or bombs are optimised to damage aircraft or vehicles and the preferred weight is in the range 5–10 grammes. The larger the velocity, but the thinner the cloud of fragments is, the less is the chance of hitting the target. Taking, then, the size of the randomly-formed fragments and the velocity to be the most important parameters, they both depend to a large extent on the ratio of charge weight to case weight. Provided this ratio is not far from 1:1, the size of the fragments decreases as the detonation pressure of the explosive increases. For moderately or very heavy cases the fragment velocity tends to be a function more of explosive work capacity and charge/case weight than of detonation pressure. In HE shells, which have c:w ratios of about 1:10, velocities average 1000 $m\ s^{-1}$, while large bombs, with ratios up to 2:3, can give velocities up to 4000 $m\ s^{-1}$. Fragments, being of poor ballistic shape, decelerate more sharply than bullets: a typical fragment slows by more than half over a distance of 100 metres. It will require a velocity of some 1000 m s^{-1}, depending on its shape, to penetrate 10 millimetres of steel. The effect of fragment shape on its penetration capacity is less at higher velocities, and at 2000 m s^{-1} it is said to be negligible.

Scabbing

Fragmentation is one of the ways in which high explosive shells are designed to cause damage: another is scabbing. Scabbing is a technique for damaging a thick metal plate, like armour, on one surface by exploding a charge on the opposite surface. The charge must be detonated from a point furthest from the metal, whereupon the shock front meets the metal in a more or less parallel plane and passes through it as a compression wave. When it encounters the other surface of the metal it is reflected, because of the large acoustic difference between metal and air. The wave undergoes a phase change on reflection, and travels back as a tension wave. Once the tension wave has cleared the later portion of the compression wave the momentum which has been imparted to the reflecting surface takes effect and tears the metal apart at the weakest plane, which is that of the returning tension wave. Thus a disc of metal, formed into a shallow dome and called a scab, becomes detached from the surface of the plate and is thrown forward at a speed of about 100 m s^{-1}. The diameter of the scab is roughly equal to that of the charge producing it. The military application of scabbing is in the defeat of thick armour plate on tanks. The vehicle is attacked with a type of shell known as High Explosive Squash Head (HESH). It has thin walls, a thin rounded nose and a base fuze. The shell is filled with a soft, insensitive mixture of high explosive and wax, and when it strikes the target the nose and walls deform under the impact and the explosive becomes momentarily piled against the armour. After an optimal delay of the order of a millisecond the fuze detonates the main filling and the resulting scab, weighing several kilogrammes, ricochets round the interior of the tank, causing sufficient damage and injury to put the vehicle out of action. The filling of a squash-head shell must be of high detonation pressure as well as low sensitiveness. A mixture of RDX and wax in the appropriate proportions 90:10 is usual.

Rock Blasting

The same capacity of an explosive to impart a high pressure shock wave to its surroundings, and the ability of rock, like metal, to reflect the wave at an air interface, is vital to the blasting of rock by high explosives. Routine blasting in quarries is called bench blasting, the bench being an artificial cliff within 20° of vertical. Holes are drilled from the top surface, parallel to the face and ending at or just below floor level. Each hole is filled to about three-quarters of its length with explosive, initiated either by a detonator or by a detonating fuze. The remainder of the hole is stemmed with clay, sand or gravel. In practice a row of such charges is detonated together, but we will consider one hole for simplicity. The shock wave starts from the initiator and travels down the hole, spreading radially into the surrounding rock. Since the wave front is roughly conical, pressure is exerted in both radial and axial directions. When the wave has moved clear of the hole, these pressures are relieved unevenly and the rock, which is much weaker in tension than in compression, begins to crack in radial orientations, starting a little outward of the hole. The crack velocity is probably about 1000 m s^{-1} whereas the shock wave is travelling several times faster. That portion of the spreading wave which meets the cliff surface, often called the free face, is reflected back as a tension wave, and on its return journey meets the cracks which are spreading both away from the hole and towards it. The interaction of the tension wave with those cracks which are nearest to parallelity with it causes further reflection and cracking. As a result, the whole rock mass within a 90° sector of the hole breaks up. The residual pressure of the hot gas in the hole heaves the mass forward a little way before it collapses to the quarry floor. The whole process is a combination of shock wave effect and the work capacity of the explosive. If the burden (the distance between the hole and the free face) has been judged correctly, and the hole is properly stemmed, a minimum of the available energy is wasted as heat, air blast and ground vibration. The explosive used must be chosen for its characteristic velocity, and hence detonation pressure and brisance, to match the strength of the rock and break it into the optimum size for mechanical handling. Those pieces which are too big for this are subjected to secondary blasting, one type of secondary blasting is known as plaster shooting. A small slab of high velocity explosive or plaster gelatine, is laid on the boulder, covered with mud and detonated. The interacting internal reflections of the shock wave shatter the boulder. This is a shock wave effect only, the internal energy of the gas is largely wasted.

Elementary Shock Wave Shaping

The damage capacity of an uncased explosive charge comprises two effects, the detonation shock wave and the expansion of the gases. Both of these effects rely fundamentally on the chemical energy which is the heat of explosion of the charge, and since no practical high explosives of significantly greater energy are ever likely to be discovered, much of the research into explosives is devoted to making the best use of the energy available. In particular, the detonation wave which travels through and away from the main charge can be suitably shaped to produce a concentration of energy in a particular direction. This section discusses the shape of detonation waves and the ways in which they can be influenced.

Explosive charges of many sorts and sizes tend to be cylindrical in shape for various reasons (e.g. ballistic shapes, drilled shot-holes, filled tubes) and they are often, though not always, detonated from one end. In a cylindrical charge of very large diameter and initiated from one end, the detonation wave would theoretically, spread spherically from the detonator and would therefore approach planarity as it proceeded along the charge. In practice, however, heat losses at the circumference cause the wave front to reach a constant state of convexity (Figure 2.2).

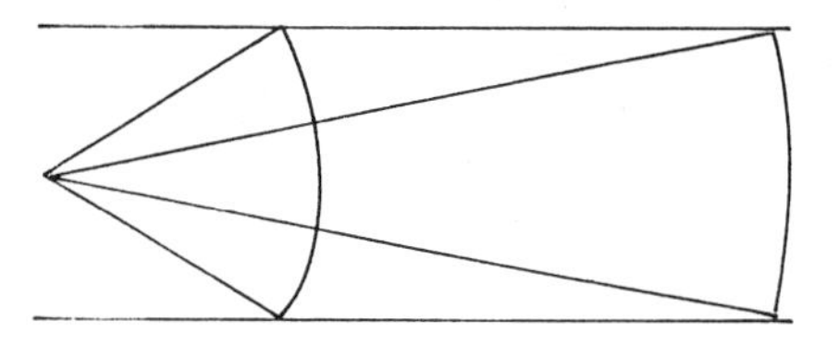

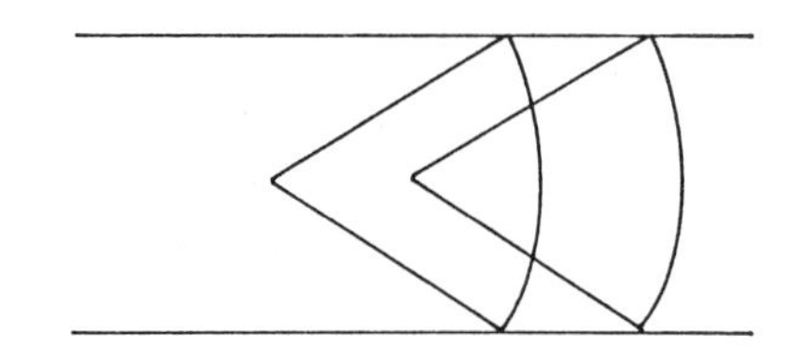

FIG. 2.2 Theoretical and actual shapes of detonation shock waves

There are two stages at which the convexity of the wave front can be modified. The first is during its progress through the bulk of the charge, where an interface between two different explosive media, called an explosive lens, can be contrived. The second possible stage is when the wave leaves the outer surface of the charge, i.e. the explosive/air interface, the geometry of which can be made to alter the wave shape. Methods relating to these two stages will be referred to as internal wave shaping and surface wave shaping respectively.

Internal Wave Shaping: Explosive Lenses

The simplest case of wave shaping arises in a composite cylindrical charge consisting of two solid explosives of detonation velocities D_1 and D_2 respectively, bonded together at a planar interface whose centre is X (Figure 2.3). If $D_2 > D_1$, then when the central portion of the wave reaches X it will accelerate from D_1 to D_2 while the outer portion is still travelling at velocity D_1. The curvature will then increase as shown in Figure 2.3.

This assumes that the second explosive charge at once assumes its own characteristic velocity when initiated by the detonation wave from the first charge. In a real situation the detonation shock wave crossing the interface will decelerate or accelerate slightly depending upon the relative velocities of D_1 and D_2.

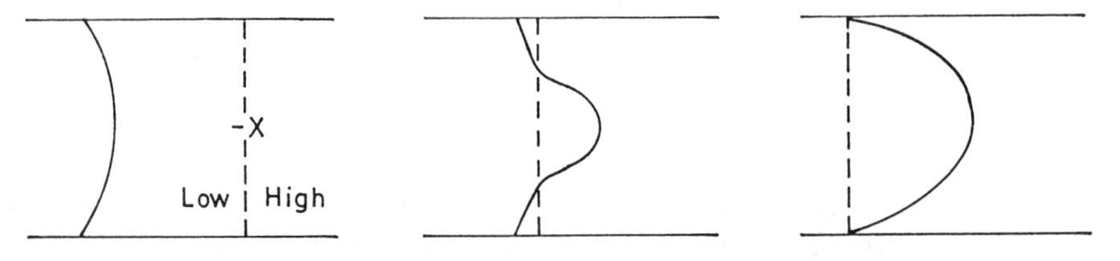

FIG. 2.3 Shockwave propagation at plane interfaces

A converse effect occurs when a wave in a steady state of curvature decelerates at a planar interface. If the ratio of the velocities is correctly matched to the initial curvature, an almost planar wave can be achieved.

Such wave-shaping effects are correspondingly greater when the interface between two explosive media is not planar but curved. In Figure 2.4 it is seen that a wave passing from a high velocity explosive to a lower velocity one via an oppositely convex interface has its own curvature reversed.

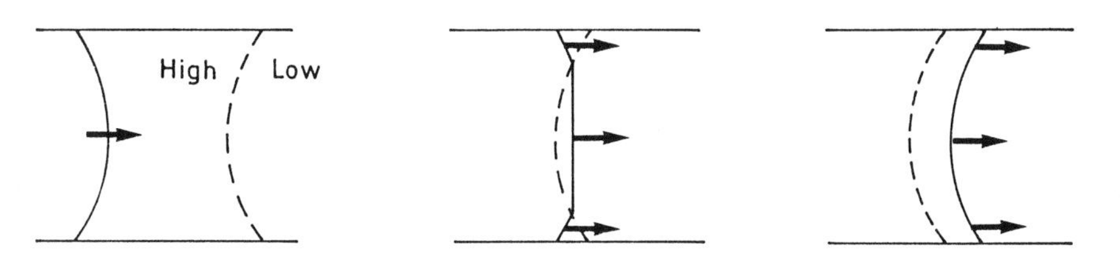

FIG. 2.4 Shockwave propagation at curved interfaces

There is an obvious analogy to the optical lens, which alters the parallel nature of a beam of light by virtue of (a) the lower velocity of light in glass compared with the air medium and (b) the curvature of at least one interface. Because of this similarity the juxtaposition of different explosives at a curved interface is called an explosive lens. The explosive lens can have a reverse arrangement to that shown above and the two types are shown together in Figure 2.5, both producing, in this case, planar waves.

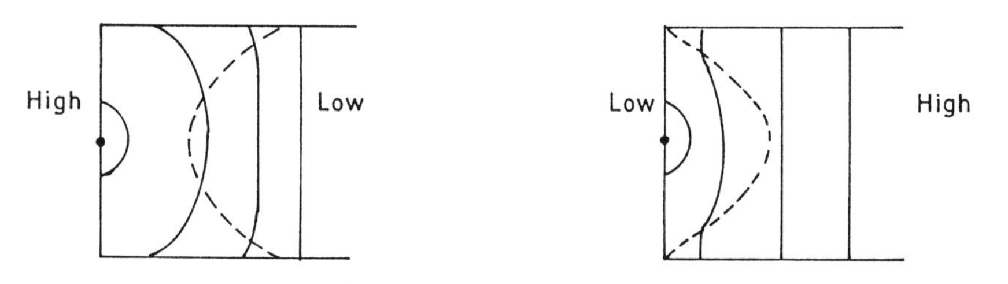

FIG. 2.5 Explosive lenses

In practice the range of detonation velocities achievable in suitable solid explosive charges lies in the range 4000–8500 metres per second. Bipartite explosive lenses constructed from baratol (D = 4200 m s^{-1}) and HMX/TNT (D = 8400 m s^{-1}) achieve a detonation velocity differential of 1:2. Differentials of 1:4 probably represent the limit achievable, such could probably be attained by using low velocity gelatines in conjunction with suitably bonded HMX.

The Implosion Device

One of the most important applications of the convergent explosive lens is in the nuclear fission bomb or warhead. The implosion device of the prototype atomic bomb was developed by G. Kistiakowsky and operated successfully in 1945 in New Mexico and at Hiroshima. It consisted of a set of twelve bipartite charges arranged round a hollow core of fissionable material. The charges were so shaped as to form a complete hollow shell of explosive (Figure 2.6). The twelve detonators were fired simultaneously and the resulting spherical wave impinged on the core to initiate the chain reaction. The implosion device remains today one of the two normal methods for triggering an atomic explosion.

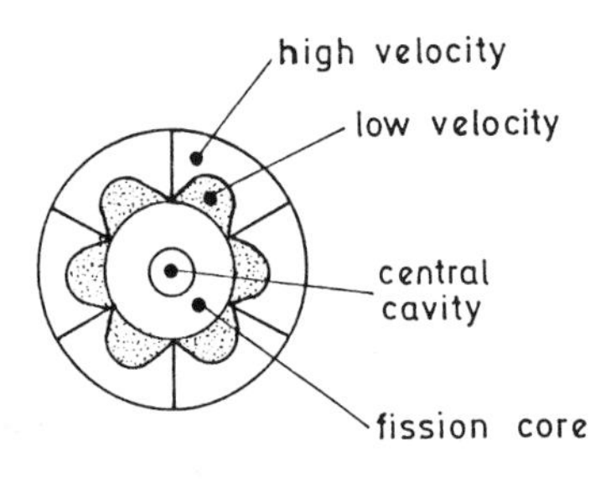

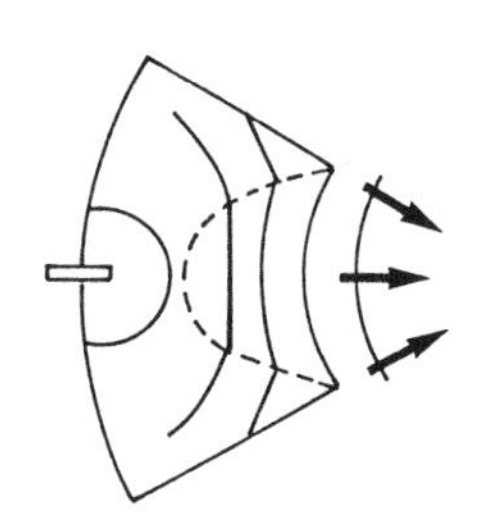

FIG. 2.6 A schematic nuclear fission bomb

Other Types of Internal Wave Shaping

INERT PAD

In Figure 2.7 is shown a method of producing a wave of approximately conical shape from the normal convex one. The introduction of an inert pad causes the wave to be propagated around the pad but not through it, thus completely altering the shape of the wave.

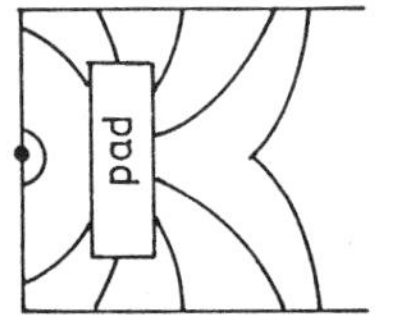

FIG. 2.7 Inert pad wave-shaper

EXPLOSIVE-METAL WAVE-SHAPER OR AIR LENS

In this system, shown in Figure 2.8, the low velocity component of the explosive lens is replaced by a layer of metal in contact with the high velocity component on the side of the approaching detonation wave and an air cavity on the other side of the liner. When the detonation wave reaches the metal liner, it is accelerated across the air gap with an overall velocity less than that of the detonation wave. An air shock will travel ahead of the liner but will be too weak to initiate detonation when it arrives at the far side of the cavity. By suitable shaping of the cavity, the desired shape of shock front can be produced. The example shown is equivalent to the explosive lens system shown in Figure 2.5.

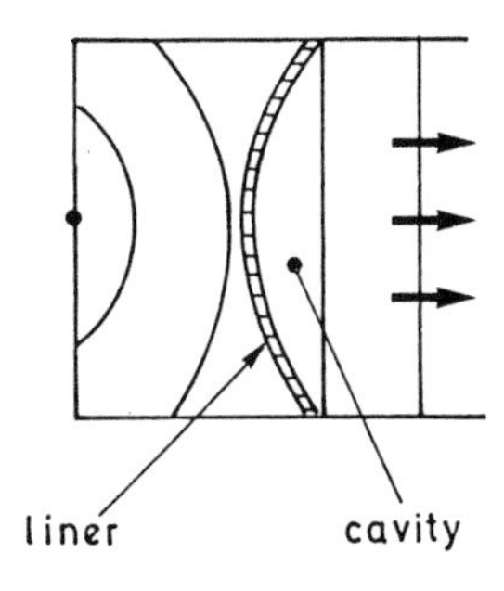

FIG. 2.8 Explosive-metal wave-shaper

CORED CHARGE

The production of bipartite charges with curved interfaces is difficult and costly. An approximately planar wave can be obtained less expensively from a cored charge, showed in Figure 2.9, in which a cylindrical core of low velocity explosive is inserted into a cylindrical charge of a higher velocity one.

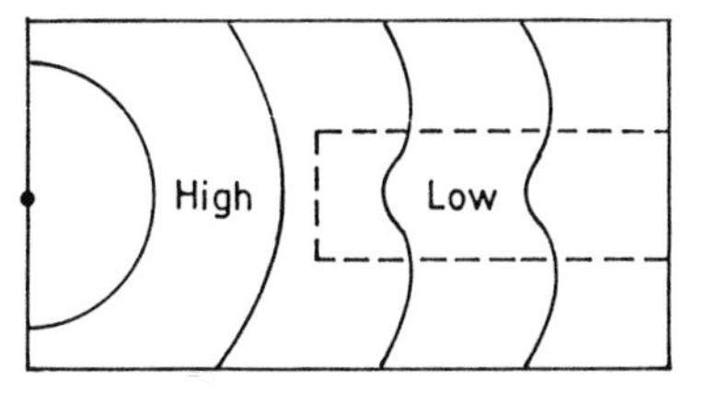

FIG. 2.9 Cored charge

Surface Wave Shaping

Thus far we have considered only the shape of the wave as it passes through the bulk of the charge. However, as has been noted, the shape of the wave is also influenced by the geometry of the outer surface of the charge which the wave ultimately reaches in its progress away from the initiation point. There is a tendency for the wave to travel in a direction perpendicular to the surface from which it emerges. Then, after the wave has passed into the surrounding air, it is no longer supported by the release of chemical energy from the charge. Therefore any edge or excrescence on the charge surface produces an immediate local attenuation of the wave in excess of the normal effect of lateral spreading.

The attenuated portion decelerates more rapidly than the remainder so the entire wave front becomes distorted. Conversely, any concavity in the charge surface produces a local strengthening of the wave which counters the normal attenuation. These effects, known as Mach disturbances, will now be considered in the case of the simple cylindrical charge and some variants.

If a cylindrical charge, such as a stick of gelignite initiated to detonation at one end, a convex detonation wave travels along the charge as we have seen. It also expands into the air with a lateral velocity component. The wave front in air assumes a pear-shaped form (Figure 2.10).

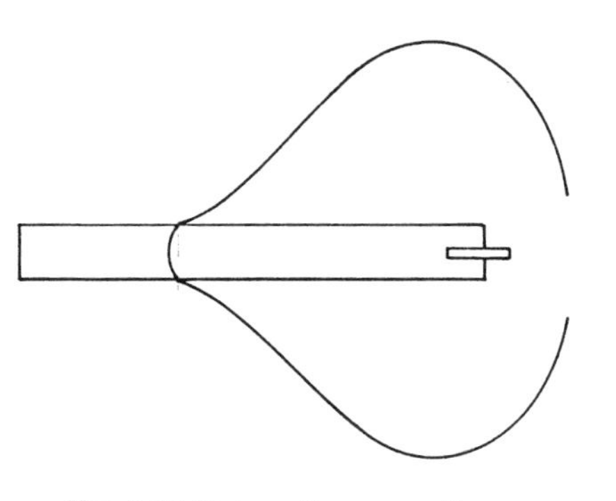

FIG. 2.10 Detonation wave from a cylindrical charge

When the wave reaches the far end of the charge and emerges completely into the surrounding air, its new shape will depend on the geometry of the end of the charge.

If the end is planar and perpendicular, the emerging wave front will contain an annular-re-entrant distortion caused by attenuation at the circular edge (Figure 2.11).

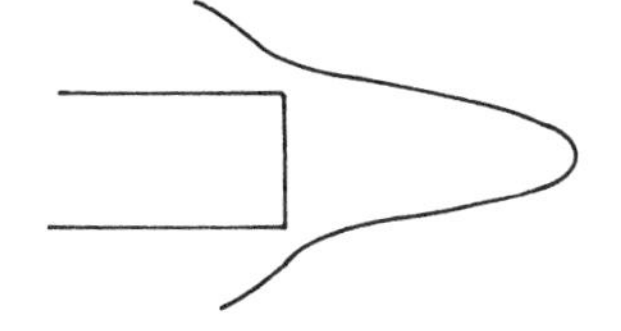

FIG. 2.11 Shockwave from a plane-ended charge

If, however, the end is hemispherical no Mach disturbances arise and the wavefront is smooth and conical as shown in Figure 2.12.

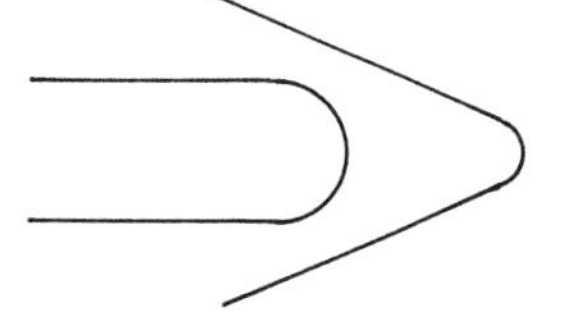

FIG. 2.12 Shockwave from a charge with a hemi-spherical end

Alternatively the end can be shaped into a wedge, in which case the wave spreads laterally from the two plane surfaces (Figure 2.13). This effect can be useful for the splitting of rock or other hard material.

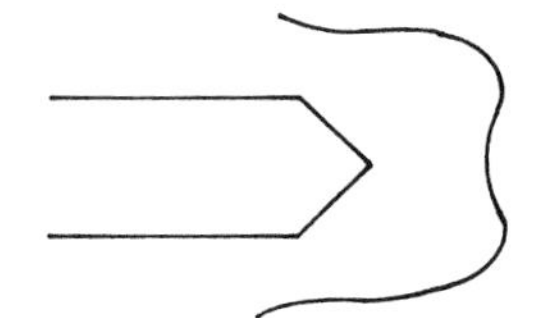

FIG. 2.13 Shockwave from a charge with a wedge

Finally, the forming of the end of the charge into a conical re-entrant shape as in Figure 2.14 produces a unique effect: the conical shock front emerging from the cavity produces a reflected wave behind it, which then becomes superimposed on the parent wave and the reinforced narrow front moves forward at great speed, followed by a jet of gas. The effect is very damaging to targets such as metal sheets, producing a localised perforation not achievable in any other way by the same quantity of explosive. The term shaped charge is sometimes used to describe this kind of charge, but since, as we have seen, the shape of any charge whatsoever affects its performance, the term cannot be recommended because it is insufficiently specific. The alternative term hollow charge though not perfectly explicit, is better and will be used herein. The hollow charge effect was named the Munroe effect after the American C. E. Munroe (1849–1938), the first man to investigate it in any scientific way. Munroe himself did not understand the phenomenon fully in 1888, which is not surprising, because even low velocity waves in detonating gas mixtures had been first discovered only a few years previously, and the study of waves in solid explosives was in its infancy. Other names connected with the hollow charge effect are Neumann, who investigated it in the period 1920–14, Lodati (1932) and Neubauer (1937). Most of the work on it, however, has been done in anonymity and great secrecy, appropriate to its immense military importance.

Hollow Charges and Linear Cutting Charges

Improvements

The uprating of Munroe's original discovery into a far more efficient penetration method was based on two vital improvements:

- ▷ The hollow in the explosive was lined with a thin layer of inert material, usually metallic. This was found to re-form into a narrow, fluid jet which has much greater penetrating effect than the gaseous jet which it largely replaces.

- ▷ In order to allow space for the jet to develop and lengthen, a gap ('stand-off') was provided between the mouth of the charge and the target. This is an opposite requirement to most of the shock effects of explosives, which rely on intimate contact between explosive and target.

These discoveries were made in time for the 1939–45 war, in which hollow charge weapons and demolition devices were used extensively. Today they are used in an even greater variety of weapons, and also in the field of commercial explosives.

Configuration

A conical, lined hollow charge typically has a length of about one cone diameter behind the apex of the cone and is invariably detonated from the rear. A single detonator on the axis is usual. The charge does not need to be cylindrical throughout its length, so it is often tapered to the rear (Figure 2.14).

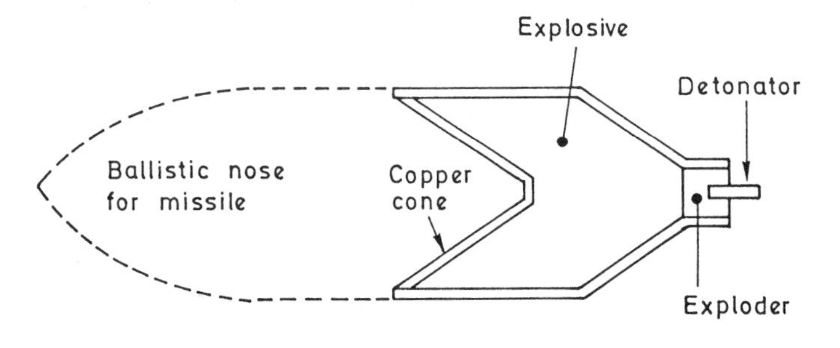

FIG. 2.14 Diagram of lined hollow charge

Formation of Jet

Some means has to be found to provide a stand-off between the mouth of the cone and the target of at least one cone diameter, and preferably about three. For a static detonation only a simple support is needed, but for a missile warhead some form of thin ballistic nose and one of several alternative fuzing systems has to be provided.

It is important that the detonation shock front reaches the apex of the cone with little convexity. It then presses the liner radially into the axis as it proceeds. Thin though the liner is (average only 2.5 per cent of its diameter), the material shears in the process, the front face moving forward to form a narrow axial jet, while the material from the rear face forms a thicker section, known as the slug or plug, near the original position of the apex. The front of the jet emerges at an extremely high velocity, up to 8000 m s^{-1}. Behind the tip there is a gradient of velocity as well as temperature, so the jet lengthens as it proceeds. Eventually it breaks up into discrete droplets of metal, and it is just before this happens that its penetrating capacity is maximal. The rearmost part of the jet, comprising material from the periphery of the cone, has a velocity of only about 1000 m s^{-1}, and behind that travels the slug at only about 300 m s^{-1}. Being less affected by the heat and pressure of the detonation wave, it remains a little below its melting point, in a state known to metallurgists as ‘pasty’.

Method of Penetration

The penetration capacity of the jet is due to its great kinetic energy directed at a small area of impact. If the jet has mass M, velocity v and cross-sectional area A, then its penetration capacity is proportional to $Mv^2/2A$ where v is very large and A is very small. The penetration of a thick metal target by the jet is not due to the melting of the target material or to erosion, but to plastic flow of the material when exposed to a pressure of some 230–300 kbar for about four microseconds. No material is lost from the hole unless the hole reaches the further surface. The hole is typically narrow, tapering, not quite straight, and for a copper liner against steel, about five cone diameters (CD) deep. (In the latest anti-tank missiles this represents a penetration of up to 50 centimetres of steel). The slug does not contribute to the penetration and is often to be found lodged in the entrance of the hole.

Variables

Because of the long history and many applications of the hollow charge, a large number of variables exists. These include

- ▷ Detonation parameters of the explosive
- ▷ Ratio of charge length: CD
- ▷ Ratio of stand-off: CD
- ▷ Ratio of liner thickness: CD
- ▷ Liner geometry
- ▷ Liner material
- ▷ Fuzing mechanism (for a missile)
- ▷ Effect of spin (for a gun projectile)
- ▷ Cost effectiveness

Some of these variables interact with others, so the optimisation of a hollow charge for a particular use needs a great deal of consideration. Taking only two factors, the detonation parameters of the explosive and cost effectiveness, it is found that penetration is a function of jet velocity and this in turn depends on two other things, detonation velocity (D) and detonation pressure (ρ). The latter depends on the former and both are related to the charge density Δ:

$$\text{detonation pressure } \rho = k \times \Delta \times D^2$$

Types of Explosives

Thus the penetration capacity of the jet, other things being equal, bears a good linear relationship to the value of ΔD^2 for the explosive employed. Now, explosives of the highest D values tend to be the most expensive. For an application where the maximum degree of penetration is essential, e.g. a precision-guided anti-tank missile, a costly HMX-based composition is a candidate, since the cost of the charge will anyway be small compared with the sum of £10,000 or more represented by the whole round. However, for a less demanding military use, e.g. a trench-digging aid or the bomblets for a cluster bomb, a cheaper RDX/TNT mixture is more cost-effective.

Commercial Uses

Hollow charges are used increasingly in commercial blasting. They vary in size from small, precision-made ones, a few centimetres in diameter and supplied complete with an adjustable support, to the large ones employed in the Nobel Offshore Pack which are used in pallets of 81 for the simultaneous blasting of an area of sea-bed. These charges, some 30 centimetres in diameter, have aluminium liners and are filled with a liquid explosive just prior to sinking and firing. Such a charge would have little effect on a thick metal target but is cost-effective in blasting mud and rock. For the use of any hollow charge under water it is necessary to exclude the water from the hollow space of the cone, so that the jet can develop. This is done by an insert of foamed plastic, or by interposing an axial container and driving out the water by means of compressed gas from a cylinder.

Plate Charges

While most axial cavities in charges are conical for maximum penetration, a hemispherical cavity can be used if a shallower but wider penetration is required. Such charges have an advantage over conical ones in that the liner can be produced in an *ad hoc* fashion by explosive forming, given a stout metal mould and a disc of copper or some other malleable metal. Once the liner has been formed and fitted into a suitable cylindrical container, it is a simple matter to pack plastic explosive behind it, thus making a surprisingly effective hollow charge. The liner material is projected in a roughly tadpole-shaped mass which is more cohesive and stable than a jet. If nothing is in its way, it will travel for hundreds of metres.

Single Plane Shaped Charges

In order to provide the deepest penetration, a cone focuses the detonation wave in two dimensions, but there are also applications for single-dimension focusing, achieved by a re-entrant wedge shape in the charge and producing a planar jet. Such a jet has a cutting effect on a hard target and is therefore useful for the demolition of concrete and steel structures, such as bridges. The simplest, most compact example is in a demolition store of the Hayrick type which consists of a rectangular metal box, the bottom of which is formed into a hollow wedge shape. The box is packed with high explosive and initiated from the top, stand-off being provided by rudimentary legs.

Cutting Charges

Another application of the same principle is in the linear cutting charge. This consists of a flexible lead-alloy tube which is A-shaped in cross-section, the triangular conduit is about 1 centimetre in internal dimension and filled with a desensitised high-velocity explosive such as RDX or PETN (Figure 2.15). The two projecting edges provide a minimal stand-off. When the tube, which can be several metres long, is placed against metal plating and detonated from one end, a linear cut is produced. There are several different makes and gauges of cutting charge, denoted by the weight of explosive per metre: this is in the region of 100 g m^{-1}. The heavier gauges can cut about 1 centimetre of steel or 2 centimetres of aluminium alloy. For the cutting of metal pipes under water, where the quick placement of the charge is important, a rigid expendable collar can be made to the outside diameter of the pipe: the collar is then cut into halves and hinged on one side. The recessed inner face contains the linear cutting charge and the detonator. A diver has only to secure the collar round the pipe and return to safety above the surface before the charge is detonated.

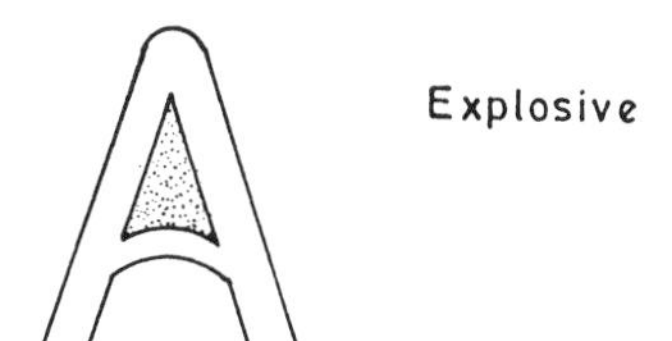

FIG. 2.15 Linear cutting charge (enlarged)

Collision of Shock Waves

We have already seen in the Dautriche method for determining detonation velocities, that detonation waves, when colliding head-on, produce a marked radial damaging effect in the plane of the collision, known colloquially as the pancake effect. Two further applications of this may be mentioned. Fracture tape is a recent innovation. It is a ladder-shaped charge of plastic explosive formed into the appropriately-shaped recesses of a mould of thin plastics material, a few centimetres wide and of any required length (Figure 2.16). In order to fracture a thick steel plate, up to say 2 centimetres thick, the fracture tape is laid upon it with the explosive touching the metal. When the tape is detonated from one end, the detonation wave travels down both sides of the ladder and branches inwards at each run in turn. The opposing waves colliding in the centre of each rung produce a rapid succession of

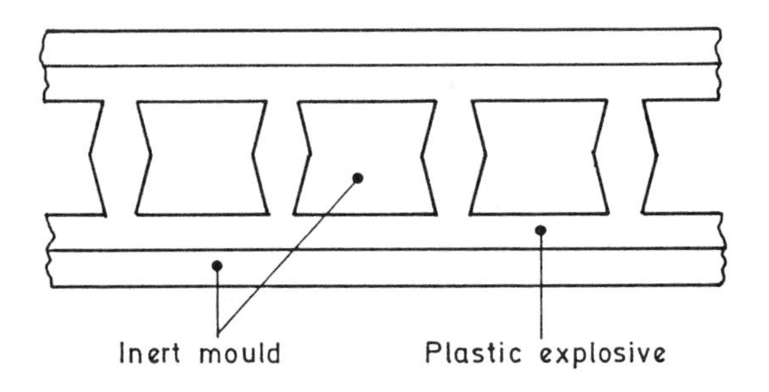

FIG. 2.16 Fracture tape

violent blows along the plate, causing it to fracture along the centre-line of the tape.

Another application of colliding shock waves is seen in the off-shore oil industry, where steel pipes, up to a metre diameter, lining shafts in the sea bed are sometimes required to be severed. A large charge in a cylindrical container is lowered down the tube to the target point and is detonated simultaneously from both ends. The collision of the shock waves at its mid-point fractures the surrounding metal more effectively than would a single-point initiation.

3. The Explosion Process: Gas Expansion Effects

In the preceding chapter we have looked at the partition of energy of explosives and considered those effects which are wholly or partly attributable to the detonation shock wave:

- ▷ Fragmentation
- ▷ Scabbing
- ▷ Hollow charge penetration
- ▷ Collision of shock waves

We now turn to the effects which are derived from the expansion of the gas behind the shock wave. They are:

- ▷ Lifting and heaving, including cratering
- ▷ Creation of airblast
- ▷ Creation of underwater bubble pulses
- ▷ Propulsion of projectiles by propellant explosives

Whereas propellant explosives do not appear in the foregoing section because they do not produce detonation effects, they will be mentioned in the following section because their work capacity is of the same order of magnitude as that of secondary high explosives even though their combustion rate is far slower.

Gas Production

Balances

We saw in Chapter 1 that an explosive is required to produce a large quantity of gas, in terms of moles per unit mass. The gas acts as a working fluid. The chemical energy of the explosive is converted into the internal energy of the gas, and hence, by virtue of the work done on projectiles of various kinds and on the surrounding atmosphere, into kinetic energy and wave energy respectively. Wave energy is a rapid alternation, for each particle of the medium, between potential energy and kinetic energy. We have seen also that for all practical explosives there is a rough inverse relationship between the amounts of heat (Q) and gas (V) produced. To some extent the balance between Q and V for a particular explosive determines its usefulness for a particular role. One reason for this is that the value of Q largely determines the temperature of explosion (T), which is an important consideration for certain roles. On explosion the energy released, assuming that none is lost in a shock wave, is apportioned between the various gaseous products according to their respective heat capacities ($\bar{c}$).

Thus:

$$T = \frac{Q}{\Sigma \bar{c}} + 298$$

where $\Sigma \bar{c}$ is the sum of the molar heat capacities of the evolved gases. These $\bar{c}$ values do not differ greatly for different gas mixtures, so very approximately it can be stated that

$$T \propto Q$$

For a large Q value and a low V value, T will be high. Conversely, if V is high and Q is low, T will also be low. Explosion temperatures are difficult to establish accurately either by calculation or experiment, due to their high magnitude and transient nature, but they vary from about 5000K for EGDN and NG down to 2300K for nitroguanidine.

For a gun propellant it is important that the explosion temperature T, or more specifically T_0 for a propellant, shall not be very high. If it is more than 2800K, problems of excessive gunflash, barrel erosion and heating of the gun arise. The corollary of a constraint on T_0 is a high production of gas (V), and this is a positive advantage for gun propellants. The chief performance parameter of a propellant is its 'force constant', the maximum amount of work which can be done by unit mass of propellant. In more practical terms, as applied to internal ballistics, it is a measure of the pressure which unit mass of propellant produces in a fixed volume, say the breech of a gun.

$$\text{Force } (F) = nRT_0 = \text{pressure} \times \text{volume}$$

where n is the number of moles of gas per unit mass (and is proportional to the parameter V as applied to high explosives), R is the molar gas constant, and T_0 is the flame temperature.

Propellant Optimisation

Since there is a constraint on T_0, and because T_0 is proportional to Q, then Q is also limited. It follows that n will be high, so a high force constant is achieved without an excessive value of T_0. In propellants for artillery and tank guns this entails the use of large proportions (up to 55 per cent) of nitroguanidine, which has the highest gas production of any common explosive (V = 1077 cm^3 g^{-1} at STP) as well as the lowest explosion temperature. In recent years certain high-performance propellants have been developed in which part of the nitroguanidine is replaced by RDX to give a higher force constant, though with some increase in combustion temperature. The other ingredients of conventional propellants, nitrocellulose and nitroglycerine, are included more for their rheological properties than for their thermochemical ones. Nitroglycerine, having a high Q value of 6275 J g^{-1} is too hot (energetic) altogether for some propellant requirements and is excluded. Nitrocellulose, on the other hand, is a doubly useful ingredient for propellants, because in addition to forming an extrudable matrix, this explosive has a chemical composition which can be varied, within limits, in the manufacturing stage and hence its values of Q and V can be chosen for a particular propellant application.

Force, Power and Strength

Release of Energy

If an explosive charge merely deflagrates, as in a gun, the release of energy is relatively slow and controllable compared with a detonation. No energy is lost in a detonation shock wave, and in a modern gun some 30 per cent of the heat of explosion of the propellant is converted into kinetic energy of the projectile. It is a law of thermodynamics that whereas all the work done in a system can be converted into heat, not all the heat can be converted into work. Therefore, looked on as a simple form of heat engine, the gun is remarkably efficient, about three times better than an internal combustion engine. High explosives, on the other hand, release their energy at a much higher rate and are much less controllable. Consequently there is a much greater wastage of energy, even under the best conditions, than in a gun. The energy is liberated in two fairly distinct ways, shock wave energy and gas expansion. In single explosive compounds of the secondary type (e.g. TNT or RDX) these mechanisms are roughly parallel properties, but for mixed explosives the same is less true. Also, as we have seen, in rock blasting the two effects work together and are not completely separable. Because of these difficult distinctions, there are several ways in which the performance of high explosives can be quantified, and this is further complicated by the different terms used in Armed Service and commercial circles.

Calculation of the Force Constant

The heat of explosion, which is the total energy contained in the charge, is also called the absolute strength. As seen in Chapter 1, it can range from 6730 J g^{-1} for EGDN to 3805 J g^{-1} for DATB, and it is even lower for primary explosives. The portion of this energy which can be utilised in doing measurable work is the sum of force × distance travelled, and is called the force constant. It is calculated as already shown and gives a much lower energy per unit mass than the heat of explosion. Thus, for example, RDX has a heat of explosion (Q) of 5130 J g^{-1} but a force constant (F) of only 1430 J g^{-1}. Although the performance of gun propellants is commonly calculated in this way, that of high explosives in a detonating mode is more usually expressed on a comparative basis and called power.

Power Index

It must be stressed that this use of the word power is completely different from the more general scientific definition, which is work done per unit time. If the latter definition were used it could be argued, for example, that a 225 g stick of dynamite of average energy would work at the rate of 30 megawatts, but that would be looking only at the rate at which the energy is released, not the rate at which it does work. The acceleration of rock or creation of airblast may take longer by two or three orders of magnitude than the detonation of the explosive, so the figure is fairly meaningless. Instead, power in the present context is the work capacity (distinguished as completely as possible from the shock wave energy) of the explosive, compared with that of a recognised standard explosive.

There are two basic ways of arriving at figures of explosive power, calculation and experiment. Taking calculation first, the method is an adaptation of the expression for the force constant, $F = nRT$. The term n, being the number of moles of gas per unit mass, is directly proportional to V, the gas volume at standard temperature and pressure (STP) for the explosive in question. Hence, working on a comparative basis, we can substitute V for n. In the same way we can substitute Q for T because the two are mutually proportional. Hence we can say that in an approximate way:

$$\text{Power} = \text{force} = nRT \propto QV.$$

The expression QV is sometimes called the characteristic product of an explosive.

In the British service the standard explosive, the power of which is assigned a figure of 100, is picric acid, for historical reasons. The Q and V values for picric acid are 3745 J g^{-1} and 780 cm g^{-1} respectively. It is easy to calculate the comparative power, or power index, of another explosive, say RDX, given its respective Q and V values as 5125 J g^{-1} and 908 cm^3 g^{-1}.

The power index is then

$$\frac{5125 \times 908 \times 100}{3745 \times 780} = 159$$

This is a percentage and is dimensionless. A set of such calculated power indices is given in Table 3.1.

Trauzl Lead Block Expansion Values

The experimental determination of the power parameter has always proved to be anomalous, due to the difficulty of separating the effects of

TABLE 3.1 POWER INDEX VALUES FOR SOME HIGH EXPLOSIVES

Secondary High Explosives	*Power Index (Calculated)*	*Primary High Explosives*	*Power Index (Calculated)*
EGDN (nitroglycol)	170	Lead styphnate	21
PETN	161	Mercury fulminate	14
HMX	160	Lead azide	13
Nitroglycerine	159		
RDX	159		
RDX/TNT 60/40	138		
DATB	132		
Pentolite 50/50	129		
Tetryl	123		
TNT	117		
Picric acid	100		

shock wave and gas expansion. The classical method, dating from the beginning of the century, used the Trauzl lead block. A cylindrical cast block of lead, 20 centimetres in height and diameter, has an axial hole 2.5 centimetres wide to below the mid-point. A 10-gramme charge of the high explosive, fitted with a prescribed detonator, is inserted in the hole, which is then stemmed with sand and the charge is detonated. The net enlargement of the hole is taken to be a measure of the explosive power. The rounded figures in Table 3.2 are for enlargements by various explosives, measured in cubic centimentres.

TABLE 3.2 LEAD BLOCK EXPANSION VALUES FOR SOME HIGH EXPLOSIVES

Blasting gelatine	520
Nitroglycerine	515
Ammonal	470
Gelatine dynamite (63.5% NG)	415
Tetryl	340
Trinitrobenzene	330
Picric acid	305
TNT	285

It is usual, however, to express these results as a percentage of the figure for picric acid, and such percentages show a reasonably close resemblance to the calculated power indexes.

The Ballistic Mortar

A different method, much more economical in time and material, is the ballistic mortar. A heavy steel mortar is suspended from a frame several metres high and a charge standardised at about 100 g is loaded into a hole tangential to the arc of swing. A substantial steel projectile is also loaded into the cavity. Upon explosion the projectile is ejected and the recoiling swing of the mortar is measured. The square of its value is an indication of the power of the explosive. This method gives somewhat different results from the lead block measurement, but is convenient for routine quality control in the manufacture of commercial explosives, provided they are nitroglycerine-based.

Commercial manufacturers make use of the mortar results to indicate to customers the performance of their products, but their terminology is rather different from that already described. Instead of the power index there are four different parameters:

- ▷ Relative weight strength
- ▷ Relative grade strength
- ▷ Cartridge strength
- ▷ Relative bulk strength

Relative Weight Strength

The first of these, relative weight strength, is the nearest to the concept of a power index, but is determined by the ballistic mortar deflection (squared) using blasting gelatine as a standard. It is defined as the strength of any weight of explosive compared with the same weight of blasting gelatine.

Relative Grade Strength

For relative grade strength a different standard explosive, straight nitroglycerine (NG) dynamite, is used and it is applied in a different way. This type of dynamite can contain various percentages of NG, so the deflection of the mortar given by a certain weight of the explosive under test is equated with the percentage NG content of dynamite which gives the same result for an equal weight. Hence relative grade strength is defined as the percentage of NG in the straight NG dynamite, which produces the same mortar deflection as an equal weight of the explosive. The anomaly in this nomenclature system is that the strengths of various grades of dynamite are not directly proportional to their NG content, therefore an explosive of relative grade strength 60 per cent is not twice as powerful as one with a grade strength of 30 per cent.

Cartridge Strength

In commercial blasting, the figure of strength per unit bulk is as important as that of strength per unit weight. The comparison of an explosive with straight NG dynamite on the basis of bulk produces an NG percentage figure called the cartridge strength. It is defined as the percentage of NG in the straight NG dynamite which produces the same mortar deflection as an equal volume of the explosive. A cartridge of given size, therefore, has the same blasting power as the same size cartridge of any other explosive of the same cartridge strength. The same anomaly of non-linearity applies as with relative grade strength.

Relative Bulk Strength

Finally, and analogous to the first parameter, relative bulk strength employs blasting gelatine as a standard but the mortar is used to test equal volume of both explosives. Relative bulk strength is defined as the strength of any volume of explosive compared with the same volume of blasting gelatine.

As a purely illustrative example, the following Table 3.3 shows what the relevant figures might be for a hypothetical explosive *X* of density 1.3 g cm^{-3}.

TABLE 3.3 EXPLOSIVE STRENGTH RELATIONSHIPS

Parameter	*Explosives in Mortar Test*	*Weight/Bulk in Mortar Test*	*Deflection[2] Linear Units*	*Results for Explosive X*
Relative Weight Strength	*X*	100 g	*70*	70%
	Blasting gelatine	100 g	*100*	
Relative Grade Strength	X	100 g	*70*	60%
	Straight dynamite 60% NG	100 g	*70*	
Cartridge Strength	X	(100 g) 77 cm^3	*70*	55%
	Straight dynamite 55% NG	(110 g) 77 cm^3	*70*	
Relative Bulk Strength	X	(86 g) 60 cm^3	*60*	60%
	Blasting gelatine	(100 g) 60 cm^3	*100*	

Energising with Aluminium

It has been known since the turn of the century that the addition of aluminium powder to explosives increases the heat of explosion (*Q*), and hence the temperature of explosion and the work capacity of the charge, whether for heaving material or creating airblast or underwater bubbles. The way in which aluminium functions is to react with the gaseous products of explosion in exothermic reactions as the following:

▷ $3CO_2(g) + 2Al(s) \rightarrow 3CO(g) + Al_2O_3(s)\ \Delta H = -741$ kJ

▷ $3H_2O)g) + 2Al(s) \rightarrow 3H_2(g) + Al_2O_3(s)\ \Delta H = -866$ kJ

▷ $3CO(g) + 2Al(s) \rightarrow 3C(s) + Al_2O_3(s)\ \Delta H = -125$ kJ

It will be noticed that in the first two reactions the gas volume remains unchanged, although it is reduced in the third, so the main result of the increased heat is a prolongation of the pressure effect. This is why aluminised explosives are frequently used when prolongations of pressure, as distinct from detonation shock wave effects, are required. There is also an enhanced incendiary effect in the atmosphere. As with most other modifications of explosives, there is a penalty: the aluminium, being a nonexplosive substance in itself, takes an appreciable time to react, and therefore reduces the velocity of detonation. For most explosives the maximum increase of work capacity is achieved with not more than 20 per cent aluminium content. A higher proportion is counter-productive, except for some military underwater compositions which may contain up to 40 per cent aluminium.

Aluminium is added to both military and commercial explosives, including TNT, RDX, HMX, water-based slurry explosives and certain types of rocket propellant. It is not added to most NG-based blasting explosives or gun propellants. Those commercial explosives to which aluminium is added contain a lower percentage of it than do their military counterparts because this additive is relatively expensive by commercial standards.

Explosive Properties Required for Particular Tasks

Shattering Materials

For the many different tasks required of explosives, it is necessary to match the properties of the explosive to the effect required. An explosive like blasting gelatine (92 per cent NG) has a high Q value, a high T value, a relatively low V value and coincidentally a high density (Δ). Because of the high T and Δ, its detonation velocity is high, hence much of the Q will be imparted to the detonation shock wave. It will therefore have high brisance, but because the gas volume V is not excessive and because some of the heat is no longer available to sustain the temperature of the gas, the heaving action required to follow up the shattering effect is somewhat limited. Hence blasting gelatine is used for breaking metal and other very hard material but not for moving large overburdens (soil, loose stones). Blasting gelatine is not used by military forces, but an analogy would be waxed high explosive used in the attack of armour by scabbing.

By contrast, an explosive such as ANFO has a medium Q value, a low T value, a high V value and, again coincidentally, a low density (Δ). These properties combine to give it a low detonation velocity, therefore little of its initial energy goes into the detonation shock wave, but instead is converted into internal energy of the gas. The main effect of the explosion will be a sustained heaving effect by the copious hot gas, and ANFO is thus well suited to the blasting of heavy burdens and overburdens. As the material hardness of the target increases, a corresponding increase of charge weight would compensate to some extent but a point would be reached where the explosive would cease to be cost-effective.

The following Table 3.4 summarises, in a simplified way, the properties required of explosives in both military and commercial applications, as far as their performance parameters are concerned. There are, of course, other properties to consider, such as sensitiveness, stability and water resistance, and these will be covered in a later chapter.

TABLE 3.4 COMPARISON OF PROPERTIES AND USES OF EXPLOSIVES

A. Detonation effect

Heat of Explosion Q^*	*Detonation Velocity* D	*Military Uses*	*Commercial Uses*
High–medium	Very high	Implosion charges for nuclear weapons, high-performance hollow charges, squash-head shells	Breaking metal
Medium	High	Fillings for HE shells, mortar bombs, grenades, fragmentation warheads, ordinary hollow charges, exploders (boosters) for for less sensitive explosives	Breaking of concrete, brick-work, hard rock, boosters for less sensitive explosives
High	Medium (heavily aluminised explosives)	Cratering charges, blast war-heads, aircraft bombs, torpedoes, sea-mines	—
Medium	Medium–low	—	Blasting of average rock and overburdens
Low	Low	—	Blasting of coal under-ground
Low	Medium (Primary explosives)	Detonators	Detonators

*Q values:

High	>5500 J g^{-1}
Medium	3500–5500 K g^{-1}
Low	<3500 J g^{-1}

D values:

Very high	>7500 m s^{-1}
High	7500–6000 m s^{-1}
Medium	6000–3500 m s^{-1}
Low	<3500 m s^{-1}

B. Deflagration effects

Heat of Explosion Q^*	*Mass Burning Rate* m	*Military Uses*	*Commercial Uses*
High–medium	Low	Rocket propellants	Rocket propellants
Medium	High	Gun propellants	Shotgun propellants
Low (Specifically, black powder)	High	Blank cartridges, primers for gun cartridges	Blasting of slate and monumental stone

* Definition of Q values is as Table 3.4A.

4. Initiation of Explosives

In modern warfare any munition must function accurately and reliably, also it must be safe in storage, handling and use. The initiation of explosives can be considered from two aspects, one being the functioning of the device, the other the potential for unintentional initiation. It must be remembered that there are two possible end effects of initiation, burning and detonation. Some explosives are only capable of undergoing the burning process whereas others may undergo either process depending on the method of initiation or prevailing conditions such as confinement. Some explosives will undergo the transition from burning to detonation and this has been considered as relevant to the safety of munitions. So the evolution of explosive devices has been towards increasing safety with high reliability. This has been achieved above all by the employment of more than one explosive material providing a chain of events leading to the correct functioning of the munition; it is called an explosive train.

Explosive Trains

The main explosive filling of a munition must not be prone to accidental initiation. To achieve this, explosive compositions are chosen which are relatively insensitive but have large energy outputs to give the required performance. Materials are known which are readily initiated by relatively small energy inputs but, in general, they have fairly low energy outputs. Also, due to their sensitiveness, it would be inadvisable to have a large quantity for safety reasons. To utilise these two types of explosive, a booster is used with properties intermediate between the two extremes. Together they make up an explosive train as illustrated in Figure 4.1.

Great safety and reliability is achieved in this way. The safety arises from the ability to break the chain either between initiator and booster or between booster and the main charge. This is equivalent to a safe versus armed situation. Virtually all explosive devices, unless they are very small, such as some pyromechanisms, will have within them an explosive train. The initiator, sometimes called primary explosive, is initiated by a small energy input, and its explosive output initiates the booster, thus amplifying the effect. The booster is chosen to be sufficiently insensitive not to be susceptible to accidental initiation yet capable of initiation by the initiator. In turn, the booster output initiates the main charge. Typically, the initiator material will be present as a fraction of a gramme in weight whereas the booster will weigh 1–50 grammes depending on the device and the main charge can be of any size desired.

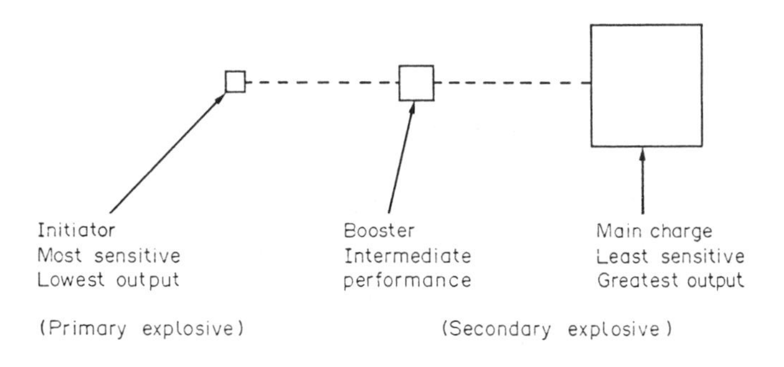

FIG. 4.1 Schematic of an explosive train

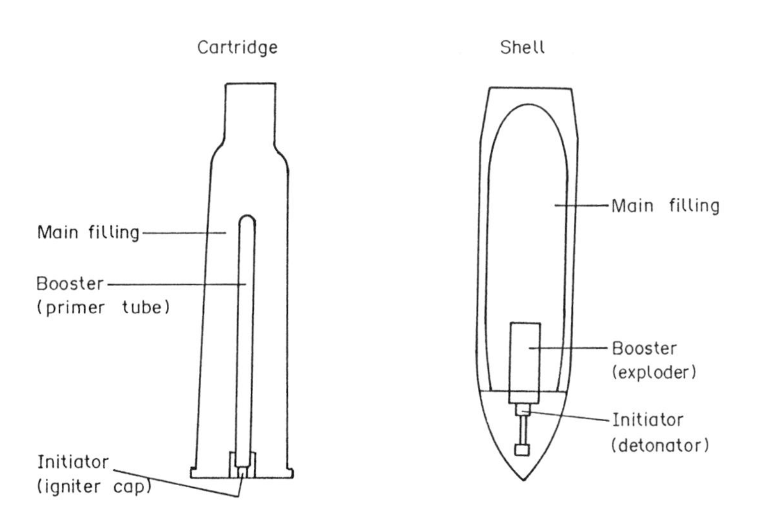

FIG. 4.2 Explosive trains in a round of gun ammunition

Explosive trains are relevant to situations in which burning or detonation is the end result. Take, for example, a round of gun ammunition, Figure 4.2. For the round to function, a stimulus is provided which initiates the primary explosive held within a cap (1) (the initiator) in the base of the cartridge. This ignites the explosive held in the primer tube (2) (booster) which flashes onto the propellant (3) (main charge). These are all igniferous events. When the shell hits a target, a primary explosive (4) (initiator) detonates. This, in turn, causes the fuze magazine (5) (booster) to detonate which detonates the main explosive filling (6) (main charge) via another booster (7) set into the main charge. In this case it is a detonation train.

Reliability is achieved because the initiator composition is quite sensitive and thus it is not difficult to provide sufficient stimulus for initiation. By choosing a suitable initiator and booster, the explosive event is carried through to the main charge.

Initiator and Booster Compositions

There are many explosive materials which are sensitive to initiation. Some are far too sensitive for practical use, such as nitrogen triiodide, much favoured for schoolboy pranks. Compounds to be used in initiator compositions must be within a sensitiveness range whereby they are not dangerously sensitive and yet readily initiated by an appropriate low energy stimulus. These exacting requirements are extremely limiting and the range of usable initiator compounds is small (Table 4.1). Also there are different types of stimuli as described below and a wide range of igniters and detonators.

To meet the requirements of a particular system it is usually not possible to use a single compound since the initiator must match the type of stimulus and the required output.

TABLE 4.1 INITIATOR COMPOUNDS

Compound	*Structure*	*Comments*
Lead azide Silver azide	$PB(N_3)_2$ $Ag(N_3)$	Detonated by all stimuli
Lead styphnate	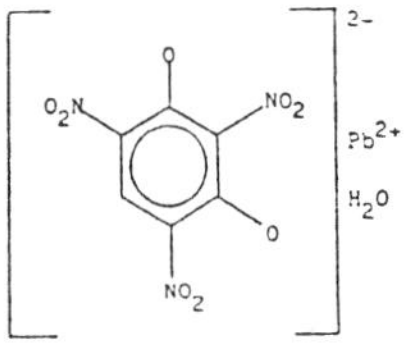 O, O_2N, NO_2, O, NO_2; $2-$, Pb^{2+}, H_2O	Normally used in igniferous initiation
Lead dinitroresorcinate (LDNR)	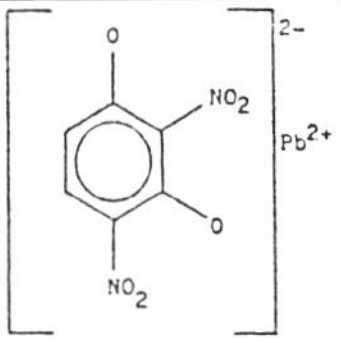 O, NO_2, O, NO_2; $2-$, Pb^{2+}	Normally used in igniferous initiation
Tetrazene	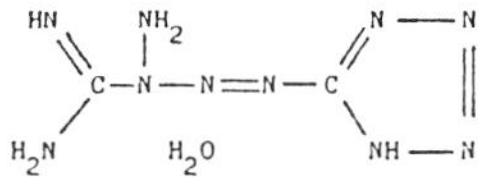 HN, NH_2, C—N—N=N—C, N—N, NH—N, H_2N, H_2O	Used to sensitize other initiator materials to stab initiation

Two methods are used to adapt the few suitable materials to provide reliable initiation. One method is to use mixtures which will optimise sensitivity and output. For example lead azide is not reliably initiated by stabbing, however, the addition of 5 per cent tetrazene lowers the required stabbing energy to around 10 per cent of that required for pure lead azide. To make compositions receptive to flash initiation, lead styphnate is employed as in ASA composition (lead azide, lead styphnate and aluminium powder). On the other hand, the composition may be modified to provide sufficient output. The von Hertz compositions, for example VH2, contain the sensitive lead styphnate to give initiation and, essentially, a pyrotechnic composition mixed in to boost the igniferous output. These complex compositions are often used when percussion is the applied stimulus as in small arms cartridges.

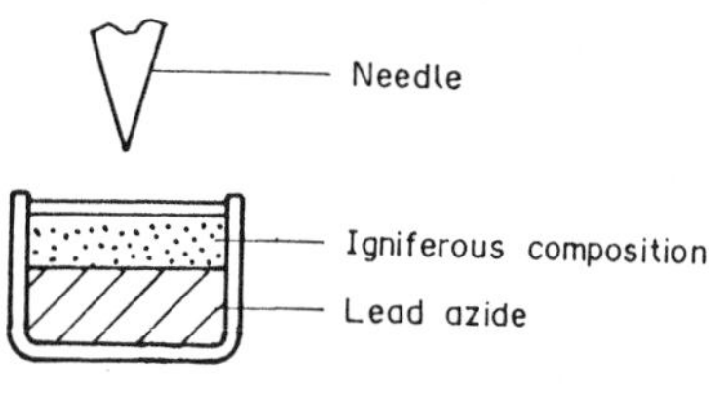

FIG. 4.3 Stab detonator for United Kingdom 81 mm mortar bomb

The other method used to employ the modest list of usable compounds is to press two or more layers of different compositions into a small container. For example, the stab detonator employed in the United Kingdom 81 mm mortar, Figure 4.3, has a stab sensitive mixture which is igniferous pressed onto a layer of lead azide which instantly burns to detonation. In many cases a booster explosive is present as a third layer particularly for detonators, Figure 4.4.

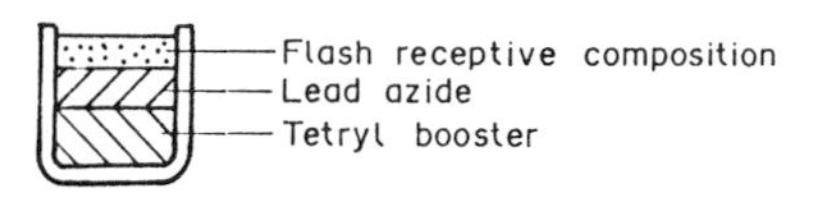

FIG. 4.4 Flash receptive detonator with integral booster explosive

Booster explosives which must have explosive properties intermediate between initiators and main charges are also limited in number. Table 4.2 lists compounds which may be used as boosters in systems which detonate. Boosters in igniferous systems usually employ either black powder (gunpowder or derivative) or fine grain propellant, both of which are readily ignited by the flash from the initiator and have good heat and gas pressure to ignite reliably a gun propellant or pyrotechnic composition.

TABLE 4.2 BOOSTER COMPOUNDS

Compound	*Structure*	*Comments*
Tetryl	CH_3, NO_2, N, O_2N, NO_2, NO_2	Losing favour due to toxicity problems
Pentaerythritol tetranitrate (PETN)	$C(CH_2ONO_2)_4$	
Trimethylene-trinitramine (RDX)	NO_2, N, H_2C, CH_2, N, N, CH_2, O_2N, NO_2	Usually desensitised slightly with wax

The Effect of Heat on Explosives

It is important to consider the effect of heat on explosives since, in the initiation of an explosive event, heat is deposited within the explosive. An optimum value of thermal stability is required for each type of explosive; initiator, booster and main charge. One method of assessing this stability is to study the effects of heating the explosive materials. As the temperature of an explosive substance is raised above ambient, a chemical reaction begins which is in fact a thermal decomposition as shown in equation 4.1. It is relatively simple to monitor this process by measuring the evolved gas and

$$\text{Explosive} \longrightarrow \text{gaseous products} + \text{heat} \qquad (4.1)$$

so the rate of decomposition can be related to temperature. A typical decomposition curve is shown in Figure 4.5. Below 100°C the curve rises gently. However, at higher temperatures, the curve steepens until the ignition temperature is reached where the reaction has reached a runaway point and an explosion usually occurs. Ignition temperatures for selected explosive compounds are given in Table 4.3. It is usual to measure these by a standard procedure since the heating rate will usually alter the point at which ignition occurs.

Low ignition temperatures certainly show compounds with low thermal stability and care should be taken to ensure that compositions containing them should not be heated, if at all possible. The nitrate esters, as a group, have relatively low ignition temperatures and thus are susceptible to accidental initiation by heat. A good example of this is the phenomenon

TABLE 4.3 Ignition Temperatures (°C) of some Explosives

Explosive	°C	Explosive	°C
Tetrazene	160	RDX	213
Tetryl	180	TNT	240
Nitrocellulose	187	Lead styphnate	250
Nitroglycerine	188	Lead azide	350
PETN	205	TATB	359

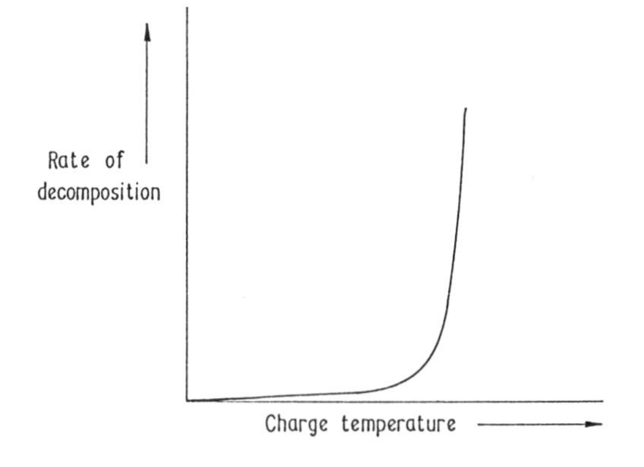

Fig. 4.5 Effect of heat on explosives

known as 'cook-off'. Gun propellants all contain nitrocellulose and some also contain nitroglycerine, both nitrate esters. If the breech of a gun becomes overheated by continuous firing, then a round put into the breech but not fired may fire spontaneously by heat transfer from the breech through the cartridge case into the propellant.

Nitramines as a group have ignition temperatures higher than nitrate esters and nitro compounds such as TNT higher still, even though it melts at around 80°C. Some explosive molecules such as (triaminotrinitrobenzene) TATB have very high ignition temperatures well in excess of 300°C and are in fact very insensitive explosives. However, lead azide, a primary explosive or initiator has an ignition temperature of 350°C. This can be rationalised when the theory of chemical reactions is considered which is a subject outside the scope of this book. A measure of ease of initiation is called the activation energy. The activation energy is often depicted by the energy curve for the decomposition of an explosive, Figure 4.6. An analogy would be a rod balanced on one end (the explosive). A push on the rod (activation energy) will cause it to topple releasing its potential energy (heat of explosion) until it ends in a stable state lying flat (gaseous products). The larger the activation energy, the more difficult to initiate the explosive.

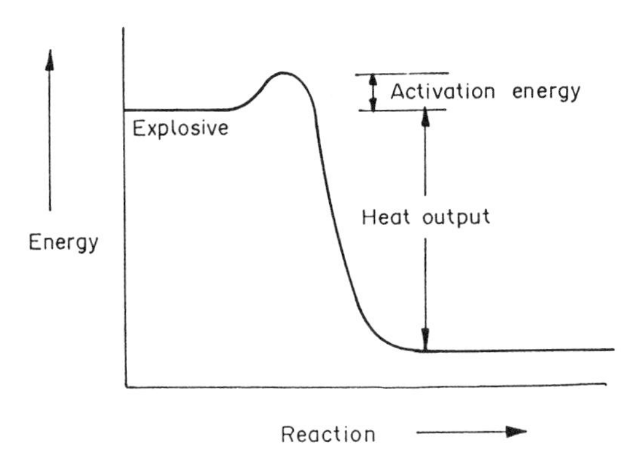

FIG. 4.6 Reaction profile of explosive decomposition showing activation energy

Methods of Initiation to Burning

Reliability and Safety

Now consider how this activation energy is provided to produce an initiation to burning. It perhaps will not be surprising that all methods, in fact, create heat within the explosive raising it to its ignition temperature. Note also from Figure 4.6 that the amount of heat evolved by the decomposing explosive is far larger than the activation energy and thus only a tiny part of the charge needs to ignite to spread throughout that charge.

As we consider below, the various energy inputs for activation, it is important to consider them from two viewpoints: the practical usefulness in a munition but also the potential for an accidental initiation. Reliability is the result of the first consideration and safety the result of the second.

External heat

We have considered in some detail, the outcome of the action of heat on explosives. It is certain that any explosive will be initiated to burning if the temperature is raised sufficiently. However, the heating of an explosive through a barrier, such as a cartridge case would certainly give unpredictable behaviour, particularly the time to ignition. This rules out external heating as a practical method of initiation.

Accidental initiation by external heating must be avoided. The cook-off of gun propellant described earlier is a good example of this type of hazard. Even less controllable would be the case of fire which could raise the temperature of filled munitions causing them to explode. We may have situations where heating of an explosive occurs intentionally, such as in a melt being poured to fill, say, shells or bombs. Great care and control is obviously required.

As the performance of weapons increases, it is more likely that some temperature increase of explosive fillings during use will occur. As missiles are built to travel at higher velocities, the aerodynamic heating will pose more of a problem and thus warheads must be protected from conducted heat. Some systems must accept the direct heating of the explosive, such as in caseless small arms ammunition, where the propellant must be in contact with the hot breech of the gun. This requires the production of a high ignition temperature propellant (HITP) as used in the Heckler and Koch G11 rifle using caseless ammunition, the first of its kind.

Flash or Flame

An obvious method of raising the temperature of an explosive material is by subjecting it to a direct flame. This flame may only be present for a mere fraction of a second but that will be sufficient to induce initiation. The potential for the accidental initiation of explosives by this method is clear and will be left to the imagination of the reader.

The practical initiation of propellants and pyrotechnics by flash or flame is a very important method; in fact the only method by which the output from the initiatior or primary explosive is passed to the booster and from the booster to main charge. A detonation train cannot be propagated by flash and thus this method is not employed to initiate the booster or main charge. However, it may be used to initiate the initiator composition within a detonation train, since some explosives will reliably burn to detonation. This will be discussed later in the chapter.

Although we are all familiar with the concept of flame, it is, in fact, a very complex entity consisting of hot gases and, usually, hot particles. When the composition used in igniters fires, there is a violent reaction which will contain, by design, hot particles. These will penetrate the surface of the next component and at the same time raise the ambient pressure. Since burning rate increases with pressure, these effects all conspire to provide reliable initiation. This is why gunpowder is such a good igniter of other explosive materials, being susceptible itself to flash initiation.

Percussion

Percussion, the striking of an explosive with a sharp blow, will generally initiate primary explosives and is a recognised method of practical initiation. Since this initiation must be caused by raising the temperature of the explosive it may not be obvious at first sight how this is achieved. Consider a diesel engine, this operates by the piston rapidly compressing air which becomes heated by a process called adiabatic compression. When diesel fuel is sprayed into this hot air it spontaneously ignites indicating that high temperatures are achieved. If the pressure can be increased rapidly, by tenfold, then many explosive compositions will be initiated. At the same time as this is occurring, particles of the composition will almost certainly rub together, creating friction and thus another source of heat. The friction effect is demonstrated by the fact that adding inert grit particles will usually make compositions more sensitive to percussion. For adiabatic compression to play its part, there must obviously be air occluded within the explosive. Thus, compacted powders will be more sensitive than cast material.

Since percussion would also equate to dropping or knocking explosives or filled munitions, it must be considered as an ever present hazard. Furthermore, if air is accidentally present in a filling, such as in the form of a crack or air pocket in a sub-standard filling of a shell, then on setback as the shell is fired the forces, say 10,000–70,000 g, will compress the gas and perhaps cause the filling to ignite. This mechanism has been branded as the culprit in some accidents where prematures, which is the explosion of a shell inside the gun barrel, have occurred.

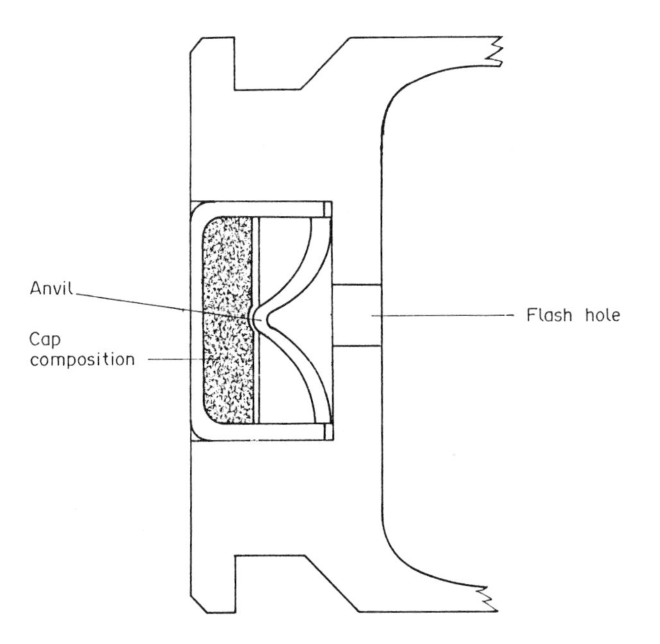

FIG. 4.7 Percussion cap ignition

Percussion is, however, a useful method for reliable initiation of an igniferous explosive train. All small arms cartridges, Figure 4.7, contain a percussion cap which is filled with a primary explosive composition, carefully chosen to be reliably initiated by the hammer of the weapon yet still allowing safe handling. The output flashes through a flash channel on to the propellant inside the cartridge case. Another example of the use of percussion caps is in most hand grenades where a spring loaded lever hits a percussion cap when the fly-off lever is released. The output in this case ignites a delay column which in turn initiates the detonation train.

Friction

The generation of heat by friction is familiar to all and it is not difficult to attain temperatures of sufficient magnitude to ignite explosive compositions. Two examples of friction initiation of explosives utilise pyrotechnic compositions; they are the household match and party crackers. Friction, however, is not particularly suitable for use in explosive devices and is generally limited to the firing of some pyrotechnic stores such as the thunderflash, which is a flash-bang training device. Friction is, however, an ever present hazard when bare explosive is present such as in manufacture and filling. Crystalline or gritty compositions are more prone to friction initiation and thus care must be taken to avoid this type of stimulus with these materials. One group of explosives which come into this category are pyrotechnic compositions which usually contain crystalline salts and fine metal powders.

Stabbing

Although direct friction caused by two rubbing surfaces is not used in an obvious manner as a common initiation method, friction must certainly play a role in the initiation of primary explosives by the process known as 'stabbing'. This is a very popular method of changing a physical stimulus into an explosive event. The stab detonator depicted in Figure 4.8 shows a typical stab initiator where a needle, usually spring-loaded, is poised above a stab-sensitive composition. When the needle is allowed to strike and penetrate the top layer of composition, frictional forces between the needle and the explosive, as well as between explosive particles themselves, create hot-spots. It may well be that a certain amount of adiabatic heating also occurs as the explosive is pushed aside. Together, these effects will reliably initiate the composition chosen for this device. As in the example given, it is a normal practice first to have an igniferous event, as with a lead dinitroresorcinate LDNR, barium nitrate and tetrazene composition. This will then ignite either an igniferous booster, if the end event is igniferous, or a composition which burns to detonation as in Figure 4.8, if the end requirement is a detonation.

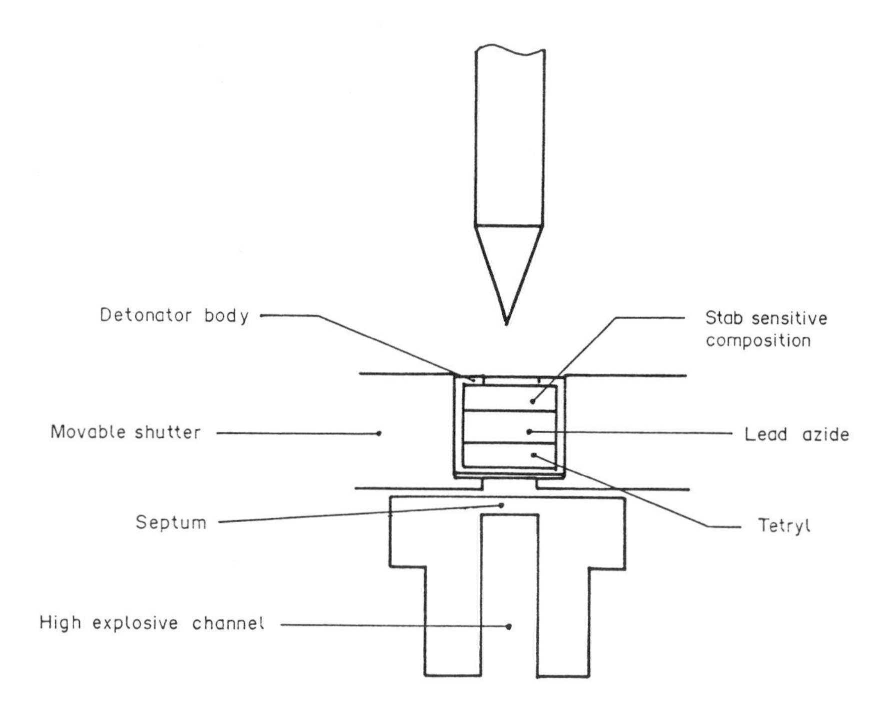

FIG. 4.8 Arrangement of a stab detonator system

Since the hot spots need only be small, in the order of 10^{-5} to 10^{-3} centimetres in diameter, the needle does not need to be large. A typical stab needle is only a few centimetres in length. The point is typically at a 20° angle. The striking velocity is required to be greater than one metre per second and penetrate to a depth of at least 1.5 millimetres. The energy required is of the order of 60 mJ. This is readily achieved by spring-loading the needle. Stab initiators are found in a wide range of munitions including shell fuzes (direct action, graze, time mechanical, time combustion and base), tank mines, anti-personnel mines, integral fuzes in anti-tank rockets, and aircraft bombs.

Electrical Systems

As a method of putting in energy, electricity would appear to be a highly accurate method. This has led to electrical initiation being one of the two most popular methods along with stabbing. The most common electrical system is the bridgewire in which a resistance wire, surrounded by an explosive composition, is part of an electrical circuit. Depending on the particular use, this device can be activated either by a continuous current being passed through the circuit or by the discharge of a capacitor. In either case the energy requirement is of the order of 10 mJ. This equates in many fuzehead type devices, as shown in Figure 4.9, to an all-fire current of between 1 and 5 amps. It is also usual to quote no-fire energy levels for fuzeheads, thus indicating what safety requirements prevail. Hazards associated with electro-explosive devices (EEDs) are discussed later.

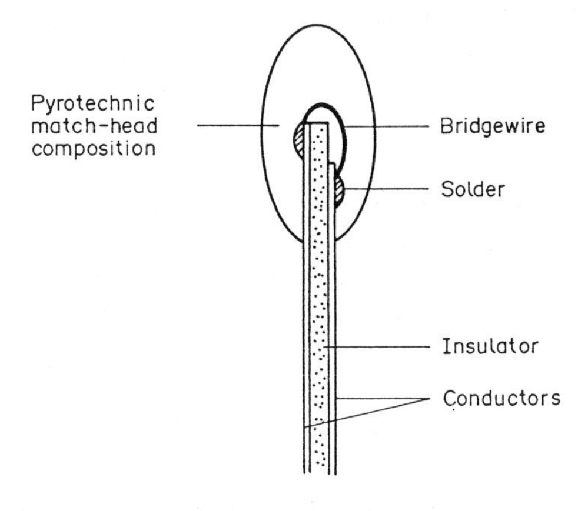

FIG. 4.9 Bridgewire initiation

Another method of using relatively low energy electrical inputs is to make the primary explosive part of the electrical circuit itself, Figure 4.10. This concept is known as conducting composition and relies on the inclusion of a conducting material such as graphite in the composition. This method has found favour particularly in the ignition of medium to large calibre guns where vehicle mounting allows a convenient source of electrical energy.

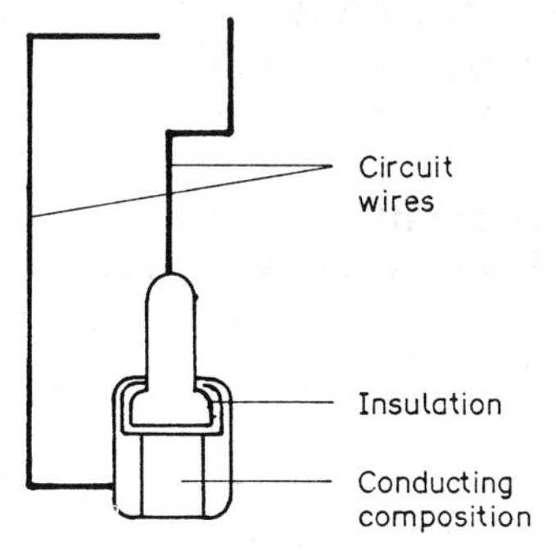

FIG. 4.10 Conducting composition initiation

Electrical Hazards

There are several instances where electrical energy could cause an accidental initiation of an explosive device or composition. Perhaps the most obvious of these is the discharge of static electricity which may range in energy from a lightning discharge, with currents as high as 300 kA supplying 10^{12} J of energy, down to the discharge of static accumulated on a person releasing up to 0.02 J. Between these extremes are static charge releases such as from moving machinery, high voltage electrical apparatus and from granular explosives themselves where friction between the grains on handling and pouring may build up a potential difference within the material.

All explosives, under all conditions, must be considered vulnerable to lightning strikes. It is estimated that the Earth receives about 100 such strikes every second. Explosives stores and magazines must be protected by earthed poles or Faraday cages to carry this energy to earth. In the open, any use of explosives where an electro-explosive initiator is being employed is vulnerable to lightning. Even a near miss may initiate an electro-explosive device (EED) even though the circuit is not complete. In commercial blasting, operations are suspended if a thunderstorm approaches within about five miles.

The lower energy static hazards need to be guarded against primarily for certain classes of highly sensitive materials such as primary explosives, nitrocellulose powders and some pyrotechnic compositions. The hazard is normally associated with handling the bare material and this is most prevalent in manufacturing and filling processes. To eliminate an accidental initiation, all buildings where these materials are handled have special earthing systems to prevent a build up of static charge on equipment or on people. Equipment is earthed directly and floors and benches are covered by a conducting material usually of a rubbery nature containing metallic powder or graphite. Workers can fasten themselves via a conducting wrist strap to the bench top thus ensuring no potential difference. Overshoes are normally worn which again are made from conducting material. Finally, the humidity is kept at a high level, usually above 65 per cent RH, since dry air encourages the formation of static build-up.

Although not strictly relevant to types of initiation, there is one other electrical hazard which should be noted. This is the phenomenon covered by the general name Radhaz, that is, current induced into the circuit of an EED by electromagnetic radiation. The actual initiation takes place by current passing through the bridgewire or conducting composition just as it would be when correctly fired. However, this current is inadvertently produced and thus a hazard. Many things may induce current and the science behind this is beyond the scope of this book. Care should be taken if EEDs are used near powerlines or powerful radio, radar or microwave transmitters. All of these are capable of inducing current in a wire particularly if the wire length is matched to the frequency. For a 300 megaherz radar transmitter of mean power 1 kilowatt the minimum safety distance has been estimated as 300 metres for the use of EEDs.

Chemical Reaction

The only modern use of an exothermic chemical reaction as a practical initiation method is in the hypergolic ignition of a liquid bi-propellant rocket motor. This is simply the spontaneous ignition of the system occurring when the liquid fuel and liquid oxidiser are allowed to mix within the combustion chamber. An example of this is unsymmetrical dimethylhydrazine UDMH reacting with red fuming nitric acid (RFNA) as used to power the American battlefield rocket Lance.

There are many examples in the chemical literature of so-called spontaneous reactions, such as aluminium powder and iodine moistened with water, although very few of these are of practical interest. There have been reports of devices initiated by the reaction of concentrated acid on a sugar-potassium chlorate mixture and by the action of water on calcium phosphide.

Methods of Initiation to Detonation

Burning to Detonation

There are three methods of initiating a detonation. The first of these has already been mentioned, that is, choosing a composition which burns to detonation. As we have seen it is not difficult to provide a burning stimulus which can be designed to impinge onto the composition which can then burn to detonation. The most common explosive which will undergo this process in a confined or unconfined space is lead azide or compositions of it. The transition is almost immediate and only small pellets are required to provide a reliable detonation. This has allowed the development of miniature detonators with cylindrical bodies 10 millimetres in length and 5 millimetres in diameter. Figure 4.8 shows a typical layered device where the burning top layer initiates the azide to burning which then detonates. Another example of this is the flash receptive azide-styphnate-aluminium (ASA) composition in the UK L2A1 electric demolition detonator, Figure 4.11, which is initiated by the flash from the fuzehead.

Burning to detonation must also be considered as a hazard, since a high order detonation is potentially more damaging and far reaching than a rapid burning or deflagration. The rate of burning r is related to pressure p by the simple equation as shown, equation 4.4, where α is the pressure

$$r = p^{\alpha} \tag{4.4}$$

index. If α is less than 1 the increase in r as p increases is somewhat self limiting (Figure 4.12(a)) whereas if α is greater than 1 (Figure 4.12(b)), there

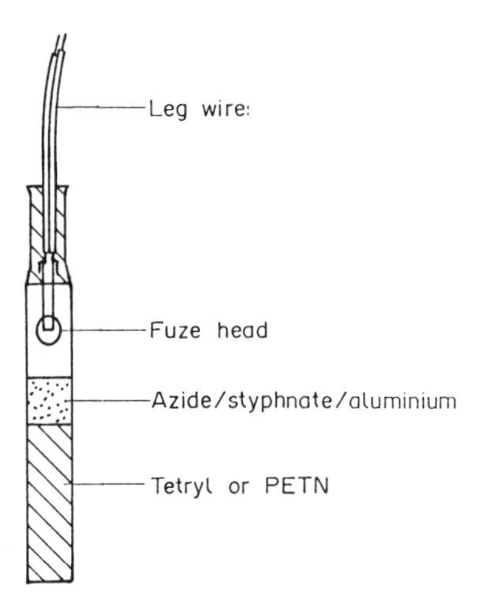

FIG. 4.11 Electric demolition detonator

is potential for the rate to increase so rapidly that a transition to detonation will occur. Thus, if an explosive composition has an α value > 1 and is ignited in a confined space, there is potential for a high order event. This may occur, therefore, when a munition is exposed to accidental heating as described earlier. Primary or initiator explosives are purposefully initiated to burn to detonation but secondary explosives are not.

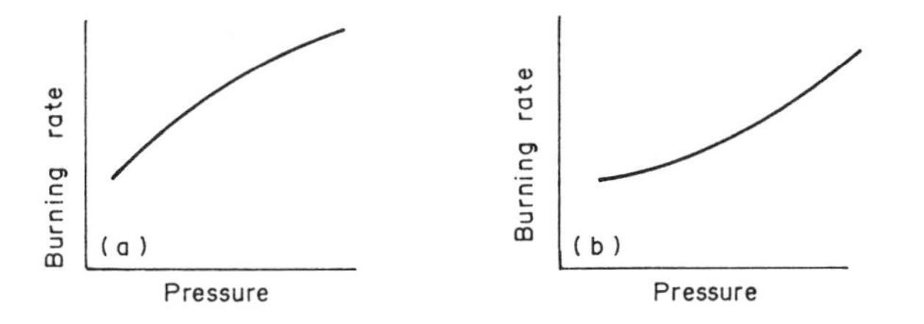

FIG. 4.12 Burning rate versus pressure

Transfer of Detonation Wave

The initiation of the booster and main charge of a detonation train occurs by the action of the detonation shockwave from the adjacent component as it detonates. This is the second method of initiation to detonation. The passage of the detonation shockwave through a charge causing propagation of the detonation has been described in Chapter 2. If this shockwave at great pressure leaves one charge (the donor) and impinges directly on another (the acceptor), (Figure 4.13), then at first the pressure will drop. However, decomposition will occur and usually the pressure will then rise until it reaches the value for a stable detonation. If the shock pressure is reduced before entering the acceptor by interposing a barrier (Figure 4.14), then the acceptor may or may not detonate depending on the magnitude of the initial pressure from the donor, the thickness of the barrier or the ease of initiation of the acceptor. Thus the acceptor with detonation pressure c (dotted line) is difficult to initiate. Explosive a, as donor, has a sufficiently high shockpressure so that reduction with the barrier is not so great as to prevent initiation. Explosion b's shockpressure, after reduction by the barrier, is insufficient to promote detonation. Explosive d (solid line), being easy to initiate will be detonated by either a or b through the barrier. Thus close contact is important in a detonation train, if it is required to function whereas an adequate barrier is required if it is not.

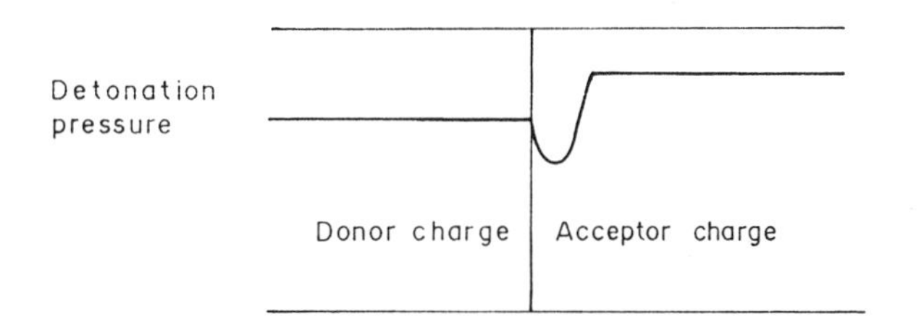

FIG. 4.13 Profile of detonation pressure for donor and acceptor charges

The third method of initiation to detonation is a special device produced originally to give an EED which was immune from the accidental initiation of induced currents. The device is the exploding bridgewire or EBW which utilises high electrical energy for initiation requiring appropriate firing circuits and energy source. The usual method is to use a capacitor which on discharge over a fraction of a microsecond produces currents of the order of 700 amps. This current, if passed through a resistance, the bridgewire, will cause it to vaporise creating a shockwave. The shockwave is of sufficient pressure to directly detonate a suitable secondary high explosive. A popular choice is low density PETN which is a secondary explosive, far less sensitive than any primary explosive. The bridgewire is set into the explosive and this device is suitable to be left in line with the rest of the detonation train without fear of accidental initiation.

A recent development akin to the EBW is a device whereby the electrical energy release breaks off a small piece of brittle plastic. This then impinges at high velocity on the explosive after a flight of a few millimetres and detonates the explosive directly. The secondary explosive of choice is either hexanitrostilbene (HNS) or low density PETN. Because of their method of action, these are usually called slapper detonators.

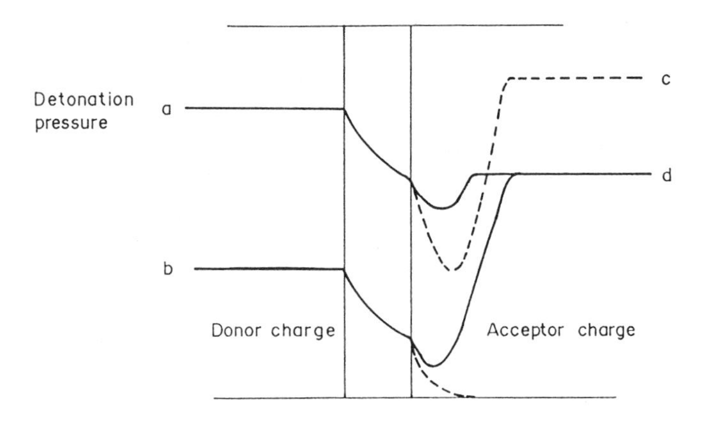

FIG. 4.14 Profile of detonation pressure for donor and acceptor with inert barrier

Safety Tests

Explosives must be capable of initiation to rapid decomposition, so there is inherent danger in their use. Obviously some materials will be more sensitive than others, for instance primary explosives. However, it is important to know exactly how sensitive all explosives are to the various stimuli described above. Tests have been evolved which fall broadly into two categories: powder tests relevant mainly to manufacture and filling processes, and charge tests to assess the vulnerability of larger charges particularly in filled munitions.

Powder Tests

These tests are carried out to assess six possible initiation situations: impact, friction, direct heat, naked flame, flash and spark. Countries which manufacture or fill explosives will have tests to assess some if not all of these stimuli. Impact testing is often the first test performed on a new explosive and there are several types of apparatus all working on a similar principle. A few milligrams of a sample, is contained below a piston or drift. A weight is dropped from varying heights until the susceptibility to initiation is assessed. In the United States this is called the Fallhammer Test and quotes fall height or energy required to give at least one initiation in ten shots. There is a German modification of this device. In the United Kingdom the Rotter Impact Test is favoured. This is possibly a more rigorous test whereby at least 200 shots of the sample and a standard are attempted. This gives a Figure of Insensitiveness (F of I) related to the RDX standard of 80. Table 4.4 shows the F of I and Fallhammer energies for a representative sample of explosives.

TABLE 4.4 RESULTS OF IMPACT AND FRICTION TESTS

Explosive	*Fall energy kp m (N m)*	*Figure of Insensitiveness*	*Pistil load kp*
Lead azide	0.75 (7.5)	30	0.01
Lead styphnate	1.5 (15)	20	0.15
PETN	0.3 (3)	50	6
Tetryl	0.3 (3)	90–100	36
RDX	0.75 (7.5)	80	12
TNT	1.5 (15)	150	36

Friction is a potential hazard and there are several tests worldwide to assess friction sensitiveness. All have some method of rubbing together two surfaces with explosive trapped between. In the United Kingdom the Mallet Friction Test was used whereby the explosive was placed on an anvil of chosen hardness made of steel, brass, aluminium bronze, Yorkstone, hardwood or softwood and given a glancing blow with a mallet of steel or wood. Finding the combination where ignition occurred gave a measure of sensitiveness. More recently, a new development has provided a more quantitative test using a rotating drum. In America the Friction Pendulum is employed whereby a steel or fibre shoe on a steel rod is allowed to swing down onto an explosive sample on a flat surface. A more quantifiable test is from Germany where a porcelain pistil is weighted on top of the sample placed on a porcelain plate. The plate is moved to and fro and the loading weight is increased in increments until an event occurs. A new sample is used for each weight change. The result is recorded as the pistil loading required to give an event and examples given in Table 4.4.

Direct heat initiation is assessed by the temperature of ignition test as discussed earlier and some results are given in Table 4.3. Sensitiveness to naked flame is tested in Britain by subjecting a sample in a 12 inch long trough, 0.5 inch diameter, to a luminous natural gas (methane) flame. The ability to support burning or a higher order event is assessed. Sensitiveness to flash can be tested by allowing a standard flash to impinge on a sample of the explosive. A suitable flash may be obtained from safety fuze containing gunpowder. Again, the type of event is noted.

Spark sensitiveness can be readily and accurately assessed by spark discharge from a capacitor through a sample of the explosive. Some materials are exceptionally sensitive to this stimulus such as lead styphnate (20 μJ) and zirconium powder of 2 micron particle size (1.5 μJ). This latter result is the zirconium-air reaction, however, all compositions containing zirconium must be considered susceptible to spark initiation.

There are also tests designed to assess the hazards associated with larger compacted charges. With these tests it is particularly important to assess the type and magnitude of event when an initiation occurs. Take the Sealed Vessel Test in which the charge is confined in a steel tube and then initiation is attempted either with an internal igniter or by external heat.

After initiation the fragments are counted and if any one test gives more than 15 fragments then that explosive is classed as a mass explosion hazard and given the United Nations Hazard Division code 1.1. If in four attempts the fragment count is always less than 10, the Hazard Division would be 1.3.

Finally, any munition must be tested for likely initiation by an accident situation and to assess its potential to create damage. The Bonfire Test assesses direct heat on the munition by placing the munition on a wood bonfire. Other tests include the Drop Test where the munition is dropped from a standard height on to a flat solid surface and the Spigot Test where the munition lands on a spike. The third important impact test is the Bullet Impact Test to assess vulnerability to projectiles. It has been clearly demonstrated that accidents in explosive storage usually spread by fragments impacting on neighbouring munitions.

5.
Gun Propellants

A gun, regardless of its size, is a device for converting the chemical energy of the propellant into kinetic energy possessed by the projectile. The gun itself, although it is essentially only a tube closed at one end, is a costly and in some ways delicate piece of precision engineering. The propellant must be carefully matched to its performance requirements, to its limitations of mechanical strength and to its resistance to erosion, so it does not degrade the effectiveness of the gun or shorten its useful life.

Conventional gun propellants consist of mixtures of one or more explosives with various additives, formulated and carefully processed to burn smoothly without detonating, under the conditions in which they are normally employed. Gun propellants are sometimes called smokeless powders, a term which originated in the 19th century to distinguish the newly-developed nitrocellulose propellants from the traditional gunpowder ones. They are indeed largely smokeless on firing, certainly so when compared with gunpowder, which gives more than 50 per cent of its weight as solid products. They are not, however, 'powders' in the ordinary sense of the word; they are produced in characteristic shapes such as flakes, ribbons, spheres, cylinders or tubes. The actual charge in a gun consists of an aggregate of such shapes, which are called grains, or if greatly elongated, sticks. Even in small arms cartridges, the individual grains are large enough for their characteristic shape to be discernible to the naked eye. The volume of each grain rises in rough proportion to the size of the gun, due to the lengthening time scale over which they are required to burn. Some typical grain shapes are shown in Figure 5.1.

In addition to the variation in size and shape, the range of gun propellants is considerably extended by the many possible combinations of ingredients which are used in their formulation.

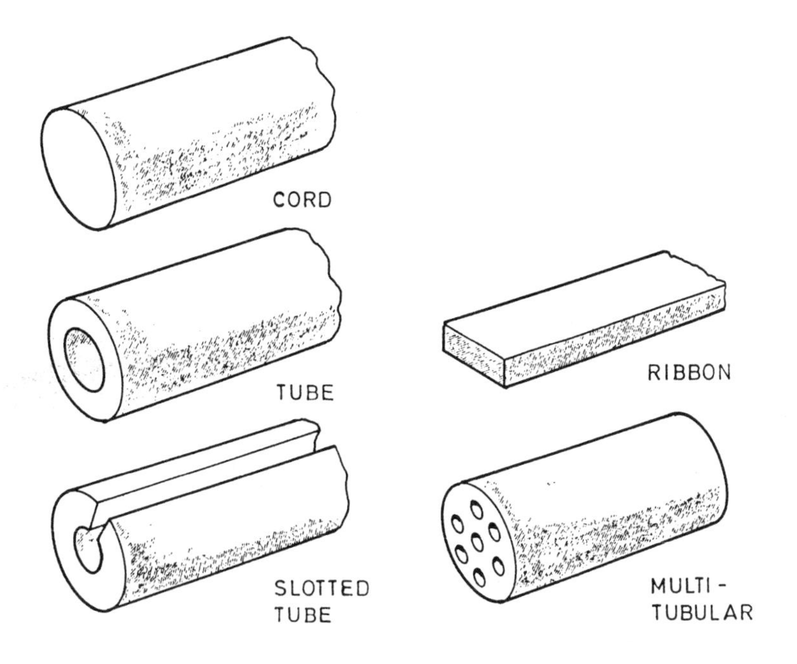

FIG. 5.1 Typical propellant grain geometries

Properties Required in a Propellant

Despite the large number of compositions which have been manufactured and tested over the years, no single formulation comes near to fulfilling all user requirements. This is not surprising when considering the summarised list below and bearing in mind that each separate requirement covers a range of problems and in some cases mutually exclusive solutions, as will be seen later. Briefly, then, an ideal propellant would possess the following properties:

- ▷ An acceptable high energy/bulk ratio.
- ▷ A predictable burning rate over a wide range of pressures.
- ▷ An acceptably low flame temperature.
- ▷ A capability of being easily and rapidly ignited.
- ▷ An acceptably low sensitiveness to all other possible causes of initiation.
- ▷ A capability of cheap, easy and rapid manufacture and blending.
- ▷ A long shelf life under all environmental conditions.
- ▷ A minimum tendency to produce flash or smoke.
- ▷ A minimum tendency to produce toxic fumes.

The Propellant Charge

The actual charge in the gun consists of aggregates of these grains, packed at the required loading density in the cartridge case or charge bag. When the gun is fired the charge is ignited by means of a primer. This device is designed to produce an explosive burst of flame, which ignites the propellant effectively within a short millisecond time scale. The flame burns rapidly over the entire surface of the charge. Propellants are non-porous, in order to provide the utmost ballistic regularity, but certain types of small arms propellant are manufactured by processes designed to produce porosity in the grains. When such a propellant is ignited, the flame penetrates the pores of the grains and so the rate of burning increases rapidly. This property is necessary in short-barrelled weapons such as pistols, and also in blank cartridges, which offer little resistance to the escape of gas. The following considerations, however, apply to the non-porous type of propellant.

The Rate of Burning (r)

Piobert (1839) showed that the burning of non-porous propellant is a surface phenomenon, in which the burning surface recedes, layer by layer, in a direction normal to the surface at every point. In the simplest case of a spherical grain, the sphere of solid propellant shrinks constantly as the flame front travels radially into the propellant (Figure 5.2). The rate of burning (r, in mm s^{-1}) is the velocity at which the flame front travels inwards at right angles to the burning surface. It can also be defined as the rate of regression of the burning surface.

The Burning Rate Law

The rate of burning (r) of a given propellant is in fact governed by the instantaneous pressure at the burning surface. The dependence of r upon P is expressed by the fundamental burning rate equation of Vieille (1893):

$$r = \beta P^{\alpha}$$

where β is the burning rate coefficient and α is the pressure index (approximately 0.9 for most gun propellants).

Though Vieille applied this equation to the burning of black powder, the relationship well describes the burning of modern propellants.

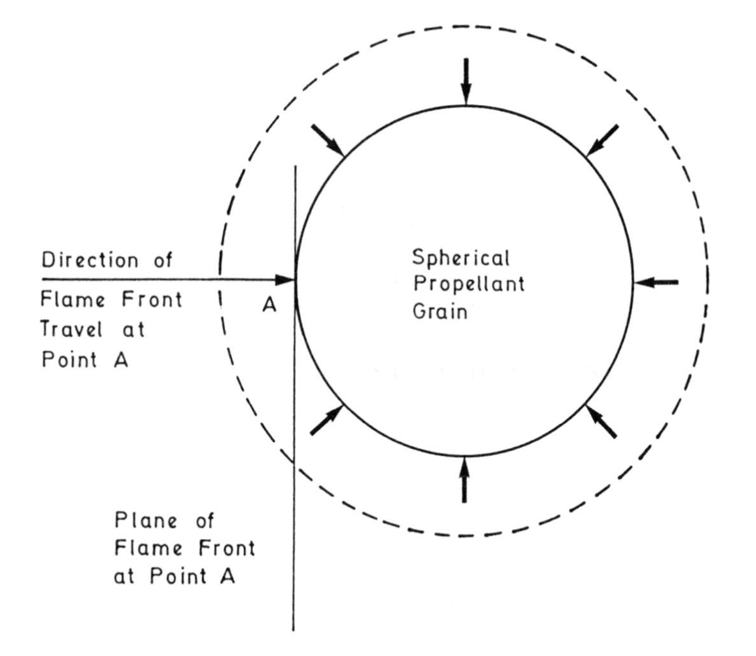

FIG. 5.2 Schematic burning of a spherical propellant grain

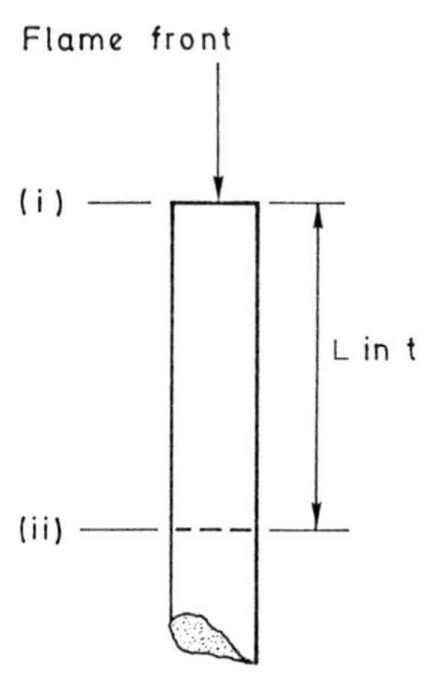

FIG. 5.3 Propellant strand

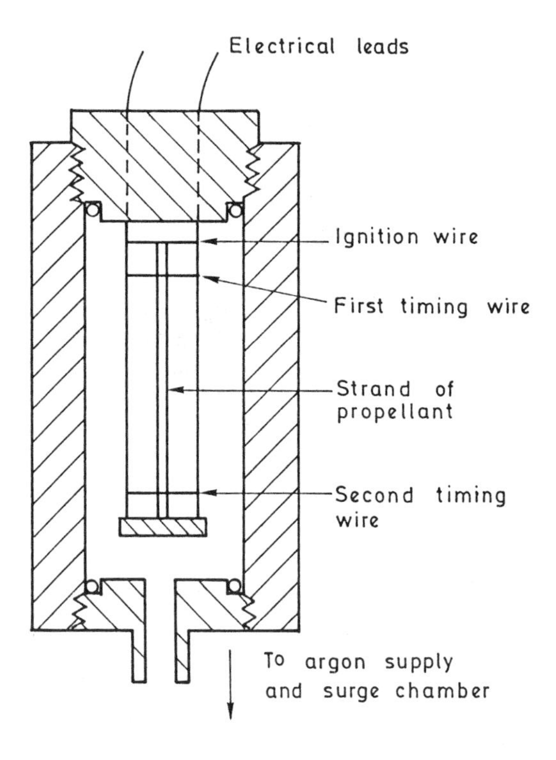

FIG. 5.4 Strand burner

The Experimental Determination of the Burning Rate

The rate of burning of a solid propellant is conveniently measured by burning a cylindrical strand of propellant at a constant pressure P. Consider the burning of such a strand according to the law of Piobert (Figure 5.3). By measuring the distance L travelled by the flame front in time t, we get:

$$r = \frac{L}{t}$$

To prevent the sides of the strand of propellant from burning, it is inhibited by coating its surface with a flame-resistant film, leaving only the end exposed. Burning rate measurements are made in a strand burner (Figure 5.4). This consists of a closed isobaric bomb filled with argon at a predetermined pressure and provided with electrical ignition and a means of measuring the time taken for the flame to travel down the strand. The propellant should be conditioned to a standard temperature (298K) before use, because the rate of burning has a small but significant temperature dependence.

The Mass Rate of Burning

The value of the burning rate r, determined by the strand burner, is required for the calculation of the mass rate of burning (dm/dt). This is the rate of consumption of a grain of propellant, or of a whole charge, in units of mass per unit time at a constant pressure P. Clearly, that volume of grain which is consumed in unit time is the product of the burning surface area (A) and the rate of regression (r). To convert volume into mass, we multiply by the density (d). Therefore the mass burning rate at time t is given by:

$$\frac{dm}{dt} = A_t \times r \times d$$

where A_t is the burning surface area at time t.

We can now adapt Vieille's equation to the mass burning rate and state:

$$\frac{dm}{dt} = kP^{\alpha}d$$

where k is a constant which takes into account the values of β, A_t and d.

When we apply the parameter dm/dt to the gun, the isobaric condition under which the fundamental parameter r was measured no longer exists. By the time the propellant is ignited, the pressure has already been raised by the explosion of the primer, and it continues to rise rapidly as the propellant releases heat and gas, until relieved by shot travel and completion of burning. The value of dm/dt will therefore also change with time, but its rate of change will be related to the original parameter r. This dependence is reflected in the shape of the pressure/time curve (Figure 5.5).

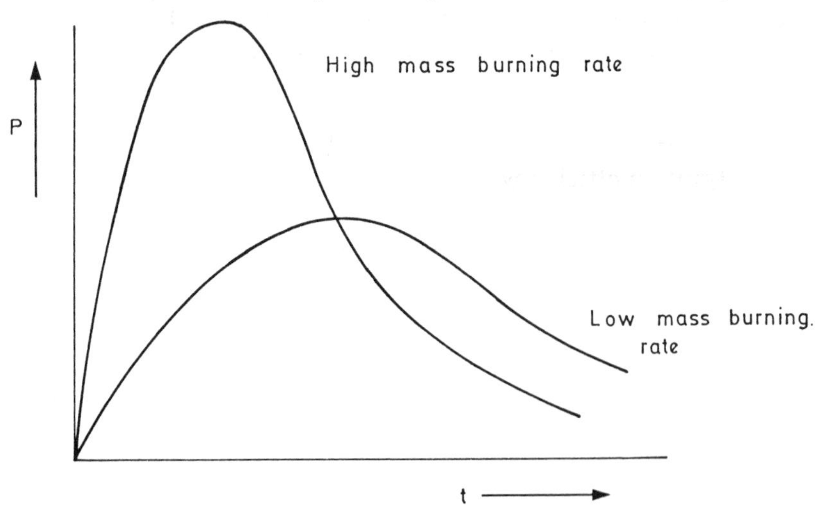

FIG. 5.5 Mass burning rate/time curve for a propellant

Grain Size and Specific Surface

For a given propellant burning at a defined pressure (P), the values of r and d are constant so that the mass burning rate is proportional to the propellant area (A).

The smaller the size of the grains of a given weight of a particular propellant, the greater will be its total surface area per unit weight. Since the propellant density is constant regardless of the grain size, then the total surface area per unit volume of propellant will increase as the grain size is reduced. The surface area per unit volume is called the specific surface of the charge and is generally measured in units of cm^{-1}.

Erosive Burning

The foregoing treatment of burning rates applies to an idealised condition known as non-erosive burning. This simplifying assumption is not strictly true. In practice, the turbulent flow of hot gases over the burning surface can erode it faster than would normal flame-front travel, and at the same time the enhancement of heat conduction and radiation at the edges causes them to be rounded off as burning proceeds. These local variations in the burning rate are called erosive burning, but in proceeding to discuss grain geometry it is necessary to ignore this phenomenon and to maintain the original assumption.

Grain Geometry

RULING FACTORS

The precise geometric form of the individual propellant grain in its unburned state is of great importance to the internal ballistics of the gun. The grain geometry governs the following factors:

- ▷ Burning surface area (A)
- ▷ Loading density
- ▷ Ignition efficiency

The dependence of A on grain geometry will now be considered.

SOLID FORM

It is evident that solid propellant grains in the form of spheres, plates or cylinders will undergo a decrease in surface area during burning. This behaviour is termed 'degressive burning', and it results in a decrease of dm/dt with time at a theoretical constant pressure.

Single-Perforated Forms

A tubular grain of propellant burns with maintenance of constant burning area, because the surface of the perforation is increasing at approximately the same rate as that at which the outer surface of the grain is decreasing. This form of grain geometry is said to have a neutral burning characteristic, resulting in a constant value of dm/dt at a theoretical constant pressure.

Multiperforated Grains

The single perforation can be replaced by a larger number of holes, up to ten or more if the grain is large enough, but as the geometry becomes more complex some extra considerations arise. For a small number of relatively large perforations, the combined effects of the enlarging inner surfaces will exceed the decrease of the outer surface and burning will therefore be progressive in its early stage. The degree of progressiveness will vary with the number and relative diameter of the holes. In all multiperforated grains, when the burning reaches the stage at which the perforations meet, the grain disintegrates into a number of longitudinal slivers. This is known as sliver-point, and the type of burning then changes from progressive to degressive. In general, a large number of perforations entails a smaller distance between burning surfaces and this causes sliver-point to occur earlier. The overall effect of increasing the number of perforations beyond, say, seven is to flatten the pressure/time curve of the charge (Figure 5.6).

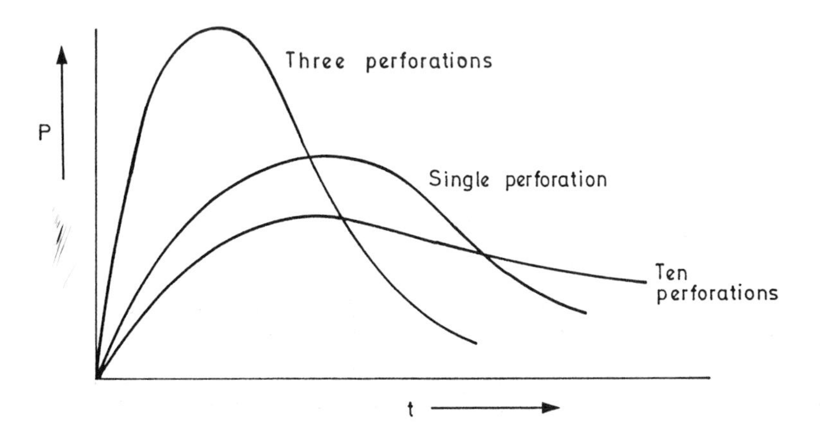

Fig. 5.6 Effect of grain geometry on the pressure time curve

SLOTTED TUBE

This is a common form of stick propellant. Although it is manufactured by extrusion as tube and slotted in the process, it burns degressively like a flat ribbon.

The Form Function

The way in which the surface area of a particular grain shape changes as it burns can be calculated and a value θ, the 'form coefficient' obtained. This has a negative value for progressive burning, a positive value for degressive burning and is zero for neutral burning. Plots of the value of the changing surface areas (expressed as a fraction of the original grain area) as a function of the fraction of the grain burnt away for grains with different θ values are shown in Figure 5.7.

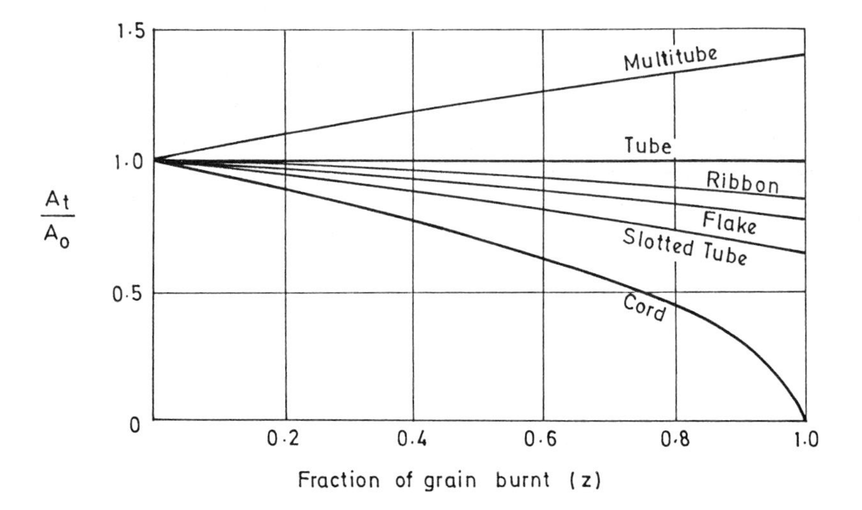

FIG. 5.7 Variation of grain area with fraction of grain burnt

Effects of Propellant Temperature on Burning Rate

As the flame front travels into the propellant during burning, each successive layer ahead of the flame is preheated by conduction, radiation and the heat from its own decomposition until its ignition temperature is reached. Obviously, the higher the initial propellant temperature, the less heat is required to raise it to its ignition temperature so the rate of burning will be higher.

The effect of charge temperature on burning rate is small but sufficient to cause an appreciable difference in gun chamber pressures and hence in muzzle velocities.

Table 5.1 shows the effects on chamber pressure and muzzle velocity for a direct fire weapon. These effects translate into differences of several hundred metres at extreme range. In use, efforts are made to protect artillery cartridges against extremes of ambient temperature, and the actual charge temperature is monitored and used to correct trajectories.

TABLE 5.1 FIRINGS IN 105 MILLIMETRES TANK GUN, USING PROOF SHOT OF WEIGHT 6.384 KILOGRAMMES

Propellant	*Charge weight*	*Ballistics at 20°C*		*Change per 10°C -40/21°C*		*Change per 10°C 21/52°C*	
	kg	*MN m^{-2}*	*m s^{-1}*	*MN m^{-2}*	*m s^{-1}*	*MN m^{-2}*	*m s^{-1}*
F527/333M	4.76	391.2	1341	6.94	11.08	4.17	5.27
NQ/M	5.68	428.0	1418	—	—	28.0	18.0

The Performance of a Propellant in a Gun

Gun propellants are required to impart kinetic energy to a projectile. The amount of energy acquired determines its muzzle velocity. If a plot of the pressure in a gun against time is made for a particular combination of gun and propellant, a curve similar to Figure 5.8 will result.

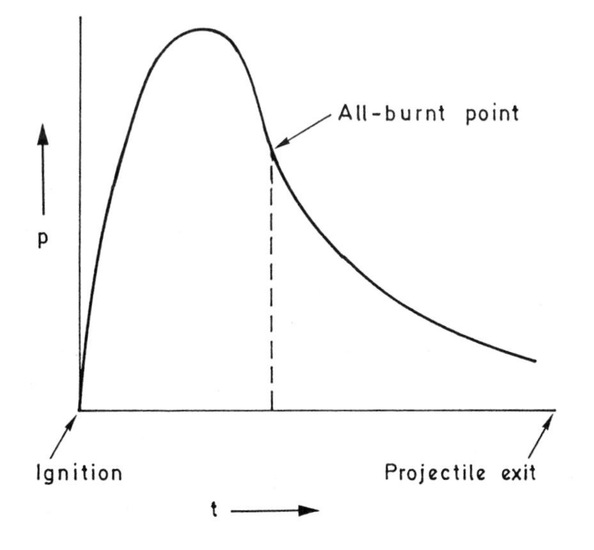

FIG. 5.8 Pressure-time profile for a propellant burning in a gun

It can be seen that initially there is a rapid pressure rise after ignition of the propellant. As the projectile begins to move up the gun tube, the volume available to the combustion product gases increases and the rate of pressure rise falls. Eventually, the volume available increases faster than the gas production rate and the pressure in the gun tube begins to fall. Soon, all the propellant is consumed (all-burnt point) and the pressure falls more quickly until eventually the projectile leaves the gun tube.

The area under the curve is a measure of the work done by the propellant on the projectile and is related to its 'Force Constant'. The shape of the curve is determined by the way in which the propellant ignites, the variation of the mass burning rate of the propellant under the changing pressure and temperature conditions in the gun, the grain size and geometry and the chemical make-up of the propellant.

This behaviour, which depends in a complex way on these various parameters is called the 'Vivacity' of the propellant. It cannot be calculated from theory and must be determined experimentally.

The shape of the pressure/time curve adopted by gun systems designers represents a compromise between conflicting requirements. On the one hand, to maximise performance and to ensure compactness of the ammunition, a high ratio of energy to bulk in the propellant is required. This entails a high force constant. On the other hand, there are limits to the maximum pressures permitted throughout the length of a gun tube due to the constraints on weight and the need for mobility. Furthermore, it is necessary to keep the all-burnt point well back in the gun tube to avoid the round-to-round variations in muzzle velocity which would result from high pressures at the muzzle. So that the curve conforms to these restraints careful optimisation of force constant and vivacity, coupled with rigorous quality control of the propellant production is necessary.

The Chemical Nature of Propellants

All solid gun propellants contain nitrocellulose. This material, being a nitrated natural polymer, gives the mechanical strength and resilience necessary to prevent breaking up of the propellant grain during handling and firing. Any such breakage would result in an altered ballistic performance of the charge. The fibrous nitrocellulose is partially soluble in some organic solvents such as acetone, ethanol, ether/ethanol mixtures and nitroglycerine. When the mixture is worked, a gel is formed which retains the physical strength derived from the polymer structure. The chemical energy of the nitrocellulose may be adjusted by varying the degree of nitration achieved during its manufacture.

Other important propellant ingredients are nitroglycerine and nitroguanidine (picrite). There are three basic types of composition called respectively single-, double-, and triple-base propellants.

Single-Base Propellant

A propellant in which nitrocellulose is the only explosive ingredient is called a single-base. It is employed in all kinds of guns from pistols to artillery weapons. The energy content (or Q value) is between 3100 and 3700 J g^{-1}.

Double-Base Propellant

A double-base propellant contains nitroglycerine (NG) in addition to nitrocellulose (NC). It is more energetic than single-base propellants and has a Q value between 3300 and 5200 J g^{-1} depending mainly on the proportion of NG present. The percentage is now much lower than when double-base propellants were first introduced in the late 19th century, due to the erosive effect of hot propellants in modern automatic weapons. Double-base compositions are now largely restricted to ammunition for pistols and sub-machine guns. The NG constituent makes them incompatible with celluloid and therefore unsuitable for use in mortar secondary charges.

Triple-Base Propellant

Triple-base propellants contain up to 55 per cent by weight of nitroguanidine as well as nitrocellulose and a limited amount of NG. Nitroguanidine, which has a pseudonym picrite, is a white crystalline solid, and it is finely ground before being mixed with the NC/NG gel to form a particulate suspension. The triple-base propellants have energy values, intermediate between those of the other two types, ranging from 3200 to 3700 J g^{-1}. Their use is restricted to larger types of guns because of their difficult ignition characteristics and because their chief virtue, the production of minimal gun-flash, is of most value in that application.

There is a further class of compositions called 'Energised Propellants' which have recently been introduced. These contain a significant proportion of the nitramine explosive, RDX.

Other Ingredients

In addition to the above compounds, all gun propellant compositions include smaller amounts of additives to impart other desirable properties. These can be classified according to their function as follows:

- ▷ Stabilisers
- ▷ Plasticisers
- ▷ Coolants
- ▷ Surface moderants
- ▷ Surface lubricants
- ▷ Flash inhibitors
- ▷ Decoppering agents
- ▷ Anti-wear additives

Some additives fulfil more than one of the functions listed above, and this convenient means of simplifying the formulation of propellants is used where possible.

There are three distinct ways in which an additive may be brought into contact with the parent composition.

- ▷ The additive can be incorporated in the propellant mix during manufacture, and thus be evenly dispersed throughout the finished product. This is the sole method applicable in the case of stabilisers, plasticisers and coolants.
- ▷ The additive can be applied to the grains after they have been formed and can then exist either as an absorbed coating (surface moderants) or a surface glaze (lubricant).
- ▷ In cartridges for larger guns, it is possible to insert the additive as a separate component in the cartridge, so that it decomposes or disintegrates under the influence of heat when the round is fired. This is an option in the case of flash inhibitors, decoppering agents and anti-wear additives.

The various classes of additive and the way in which they are used will now be described.

Stabilisers

Nitrocellulose and nitroglycerine are nitrate esters and decompose slowly at normal temperatures to form oxides of nitrogen, mainly nitrogen dioxide NO_2. If these oxides are not chemically removed from the system as they are formed they catalyse further decomposition, thus shortening the storage life of the propellant. All conventional gun propellants therefore contain an additive to neutralise the products of decomposition. The desirable properties of such stabilisers are:

- ▷ Capacity to absorb oxides of nitrogen
- ▷ Ability to neutralise acids
- ▷ Ability to form a colloid with NC and NG.
- ▷ Inertness to NC or NG before and after reaction with the decomposition products.

Some candidate compounds for use as stabilisers are mentioned below.

MINERAL JELLY (VASELINE)

This was one of the earliest stabilisers, and was employed in Mark I cordite in 1893, originally in the hope that it would be deposited in the bore by each round fired and would thus provide lubrication for the next projectile. Had this idea been successful in the way intended, mineral jelly would have been classed as what we now call an anti-wear additive. In fact, no lubricating effect was found, but it was discovered that mineral jelly, which is a mixture of the higher alkanes from petroleum, contained small amounts of alkenes which absorbed nitrogen dioxide from the propellant and thus acted as a stabiliser. Mineral jelly was therefore retained in this role for several decades and was included to the extent of 5-6 per cent in the early cordites.

Chalk

Mineral jelly does not neutralise the acids which are also products of propellant decomposition; in the early cordites this was effected by the calcium carbonate which was deposited in the mass of nitrocellulose by the process of boiling in hard water during its manufacture. In some propellants this uncertain amount of calcium carbonate was supplemented by a fixed proportion of chalk (>1 per cent) incorporated in the mix.

Diphenylamine

This compound has been used in single-base propellants from their early years to the present time. It is a white, crystalline solid which melts at 50°C. Chemically it behaves as a base and reacts with the decomposition products of NC to form nitrosodiphenylamine and some nitro-compounds. Diphenylamine is used in the single-base NRN propellant for modern 7.62 millimetre ammunition and also in the artillery propellants NH and FNH. It is too basic for use with NG and is therefore not used in double- or triple-base propellants. One disadvantage of diphenylamine is the difficulty of dispersing it evenly through the mix. Its derivative nitrodiphenylamine has been used as a superior replacement.

Carbamite

The successor of mineral jelly in double-base propellants, and later in triple-base, was carbamite. This name is a pseudonym originally coined for security reasons; the chemical name is diphenyl diethyl urea and it is known outside the United Kingdom as ethyl centralite. Carbamite forms white crystals which, when pure, melt at 72.5°C. Chemically it acts as a weak base and reacts with the decomposition products to form nitro- and nitroso-derivatives. In addition to its stabilising property it acts as a plasticiser and a coolant (see below). It is used in the small arms propellants NPP for 9 millimetre ammunition, NNN, and in the artillery propellants N and NQ.

Methyl Centralite

This is the methyl analogue of carbamite, diphenyl dimethyl urea. It has a melting point of 121°C. It is a good stabiliser and serves alternatively but not simultaneously as a surface moderant.

Plasticisers

In order to convert NC from its natural fibrous state into a gel, it must be treated with a solvent. This may either be a volatile one, later removed by evaporation, or the NG which is part of the propellant composition. In either case, the process may require the assistance of a plasticiser. The versatile substance carbamite fulfils this role well. Other plasticisers are liquid esters of a viscous nature, having high boiling points, such as dibutyl phthalate which is used in the single-base artillery propellants NH and FNH. Other liquid plasticisers are diethyl phthalate and glycerol triacetate ('triacetin'). These liquid plasticisers double as coolants.

Coolants

It is necessary to impose a limit on the flame temperature of a propellant in order to minimise erosion of the bore and other undesirable effects. This can be done by incorporating a substance in the propellant which absorbs heat when it decomposes during combustion; such an additive is called a coolant. The original stabiliser, mineral jelly, doubled in this role and so does carbamite. Dinitrotoluene represents 10 per cent by weight of the artillery propellants NH and FNH and this accounts for the conveniently low flame temperatures of these compositions. [Dinitrotoluene also imparts a moisture-repellent quality to propellants (NH = non-hygroscopic).] The three liquid plasticisers already mentioned, i.e. dibutyl phthalate, diethyl phthalate and glycerol triacetate, double as coolants and are sometimes referred to as coolant-plasticisers.

Surface Moderants

Some small arms propellants need to have their mass burning rate reduced during the early part of the combustion process, to slow the rise of chamber pressure. This is done by coating the surface of the grain with a substance which reduces the chemical energy and flame temperature of the composition. The substance is allowed to penetrate the outer layer to the required depth. The volume of the grain thus 'moderated' (or 'deterred' in American terminology) can be up to 40 per cent of its total volume. Thus the distinction between 'coolant' additives and 'surface moderants' lies not in their chemical identity but in the way in which they are incorporated in the grain; a coolant is dispersed evenly through the grain whilst a surface moderant is present only in the outer layer. Surface moderants can be either solid or liquid. Solids such as dinitrotoluene or methyl centralite must be dissolved in a suitable solvent in order to penetrate the grain, whilst liquids such as dibutyl phthalate can be applied in their natural state. Modern British small arms propellants e.g. NRN (for 7.62 millimetre ammunition) and NNN, employ methyl centralite as the surface moderant. Carbamite can be used as a surface moderant but it is more commonly employed in its stabiliser/plasticiser/coolant role.

At this point, it may be useful to summarise the possible functions of the modern additives mentioned thus far, because of the overlapping of chemical properties which is a feature of these substances.

TABLE 5.2 POSSIBLE FUNCTIONS OF SOME PROPELLANT ADDITIVES

ADDITIVE	*Stabiliser*	*Plasticiser*	*Coolant*	*Surface Moderant*	*Typical %*
Chalk	+				0.4
Diphenylamine	+				1
Carbamite	+	+	+	+	1-9
Methyl centralite	+	+	+	+	4
Dibutyl phthalate		+	+	+	2
Dinitrotoluene			+	+	10

Surface Lubricant

A glaze of graphite on the grains has been traditionally used in the case of gunpowder. In recent times, it has been applied to more modern propellants, including ballistite and NRN. The purpose of this is three-fold: it lubricates the surface to allow easy filling into cartridges; it renders the surface more resistant to moisture and it forms an electrical conductor to dissipate electrostatic charges which may otherwise accumulate.

Decoppering Agents

In a large gun, the driving band of the shell becomes partially molten during its passage along the bore and leaves a deposit of copper there. If this deposit is allowed to accumulate, it eventually effects the ballistic performance of the gun. To counter this, a decoppering agent is used with the propellant. This can be done in one of two ways. First, the propellant composition can contain compounds of lead or tin, or secondly a piece of lead or tin foil can be juxtaposed to the propellant in the cartridge. In either case, the metallic element combines with the copper deposit and forms a brittle alloy, which is removed by friction when the next projectile is fired.

Anti-Wear Additives

The erosion of the gun bore can be excessive in modern weapons unless steps are taken to control it. Coolants, as we have seen, reduce the overall temperature of the combustion gases as they are formed: an alternative is to produce a relatively cool layer of gas adjacent to the bore surface. Substances which provide this effect are known as anti-wear additives. They can either be incorporated in the propellant mix or can be a separate component of the cartridge. In the former category is a combination of titanium dioxide and talc (magnesium silicate). The latter category includes polyurethane foam liners located in the cartridge and also Swedish additive, which consists of talc incorporated in wax and wrapped in Dacron cloth.

Development of Flashless Propellants

One of the disadvantages of single and double-base propellants is that, although they are smokeless, they give a very bright flash. In World War I, the Germans overcame the problem by supplying bags of potassium chloride for use with separate loading ammunition. These were used at night-time when suppression of flash was important but, since the method gives rise to the production of smoke, they were omitted during daylight.

Flash is due to the afterburning of the hot hydrogen and carbon monoxide in the propellant gases when they mix with the air. Metal salts interrupt the free radical reactions of the burning reaction but cause an increase in smoke emission. Alternative approaches are to incorporate compounds such as picrite or RDX into the propellant which release large volumes of nitrogen on decomposition. If the flammable gases can be diluted to below their Lower Explosive Limit (4 per cent for H_2, 12.5 per cent for CO) by the nitrogen, no afterburning will occur. Typically, a triple-base propellant will contain up to 55 per cent of picrite. The effect of the picrite may be enhanced by the addition of small quantities of metal salt flash inhibitors such as potassium sulphate, potassium nitrate, potassium aluminium fluoride and sodium aluminium fluoride (potassium and sodium cryolite).

Recent Developments in Gun Propellants

Caseless small arms rounds

The main functions of a cartridge case are to grip the bullet, to contain the cap and propellant and to provide obturation. It also protects the propellant from the environment and removes unwanted heat from the gun when it is ejected. The cartridge case is however an expensive component and the necessity for it to be ejected before the next shot can be fired restricts the maximum firing rate which can be achieved. This has encouraged manufacturers to develop caseless ammunition. An example is the Heckler and Koch GII rifle with its associated caseless 4.7 millimetre ammunition.

Clearly the propellant has to be changed from its traditional form of loose grains into a strong block to grip the bullet and contain the percussion cap. It must also be strong enough to withstand the stresses of automatic handling at high rates of fire. Figure 5.9 shows the form adopted by Heckler and Koch.

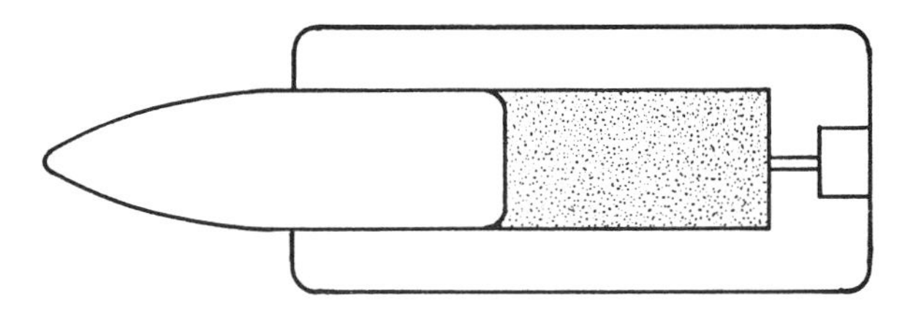

FIG. 5.9 Heckler and Koch caseless ammunition

Although the propellant must form a firm cohesive body up to the moment of firing, it must break up in a controlled way to yield an adequate specific surface and give a predictable burning rate on firing. The ignition system is therefore in two parts. The percussion cap is in the base of the round as usual and behind it is a booster explosive pellet which acts as an anvil for the cap and when ignited, gives an optimised brissant effect which shatters the rest of the propellant in a consistent way.

Since there is no brass cartridge case to remove heat from the gun, the chamber gets hot and conventional nitrocellulose based propellant would cook off. The problem has been overcome by developing a high ignition temperature propellant (HITP) which is also insensitive to accidental ignition by sparks and static electricity. Governments of several nations are displaying an interest in the concept but are keeping an open mind as to its ultimate possibilities.

Liquid Propellants for Guns

To the present time, conventional propellants for guns of all calibres have been in some solid form. Their burning has been controlled by this precise geometry. In recent years, the thermochemical performance of propellants has advanced with the inclusion of RDX in the composition. It may be said that the formulation and manufacture of solid propellants are nearing their ultimate achievable development and any further improvement in performance will come from a completely different approach.

A possible alternative approach is the use of liquid propellants, designed to be pumped directly into the chamber of the gun. Such liquids would be quite different chemically as well as physically from conventional solid propellants, and might conceivably provide higher performance than their best available solid counterparts. The use of liquids would automatically entail discarding the vast amount of expertise which has accumulated in the formulation and manufacture of solid propellants. In its place a method must be found to control the behaviour of a liquid in such a way that it provides, at the moment of combustion, the necessarily predictable surface area which solid propellants provide by virtue of their manufactured form. The solution of this problem involves a redesign of the gun even more revolutionary than that required by the caseless small arms round. Research into liquid gun propellants dates back to a German programme in World War II, by which time liquids had already been used in rockets for nearly twenty years. The Americans followed up the German work during the 1950s and there has been a resurgence of interest in the United States and elsewhere in more recent times. Some of the results are available in the open literature. At present the use of monopropellants is favoured and much research into the behaviour of hydroxylammonium nitrate has been carried out.

Amongst the various practical approaches to the problems outlined above are the regenerative and travelling charge systems outlined below.

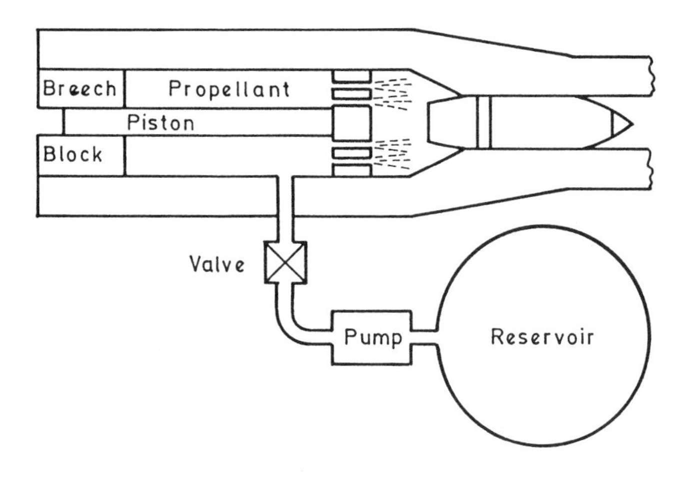

FIG. 5.10 Regenerative injection system for a liquid monopropellant

The regenerative system involves the continuous injection of propellant into the chamber during the required burning time of the gun. At an instant during the burning process there is only a small quantity of unburnt propellant in the chamber and this is present in droplet form. The injection system is to some extent analogous to a liquid-fuelled rocket motor. However, compared with the latter, there is the basic difficulty of injecting a spray into a chamber where the gas pressure is rising to a high value immediately after ignition. This problem is overcome by utilising the chamber pressure to propel a perforated piston rearwards, thus forcing the propellant forward via the perforations. This system is shown in Figure 5.10.

A variant of the above method is the travelling charge injection method (Figure 5.11). The propellant is loaded immediately behind the projectile and is backed by a rodless, perforated piston. The fuel is injected rearward into the chamber by the forward movement of the piston, and the diminishing charge is propelled forward together with the piston and projectile, which leave the muzzle together. This method is useful for a monopropellant or, by duplicating the feed, a non-hypergolic bipropellant.

The advantages which accrue from the use of liquid propellants are:

▷ Elimination of the cartridge case.
▷ Ease of manufacture, storage (in most cases) and transportation.
▷ Achievement of any specific range from a given gun elevation by precise metering of the charge.
▷ Low vulnerability to ignition by enemy fire.
▷ Easier use in armoured vehicles and under NBC conditions.
▷ Reduction of smoke, flash and bore erosion.

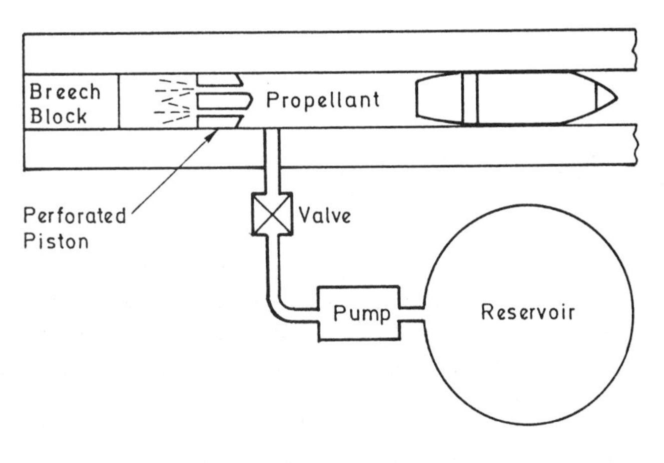

FIG. 5.11 Travelling charge injection system

After some thirty years of intermittent research into liquid gun propellants, an increased interest is being shown, since the use of metal cases is becoming economically, logistically and technically less desirable as time passes. It is believed also that certain types of liquid propellant might give better performance than that of the best solid propellants. The work done to date has shown a number of ways forward in the chemistry of candidate substances and in the revolutionary gun designs which will be necessary. However, all candidate liquid propellants have their own disadvantages and none of the possible gun designs has so far come out clearly ahead of the others. Liquid gun propellants may well come into service eventually, but probably not before the end of the decade.

Currently some 25-30 per cent of all propellant is discarded when it is removed to adjust the charge size. Liquid propellants would save this waste resulting in reduced requirements and logistic overheads for storage and transport.

6. Rocket Propellants

Rocket propellants are explosives designed to burn smoothly without risk of detonation to provide propulsive energy. Certain classes of rocket propellant are very similar in composition to the gun propellants described in the last chapter. However, due to quite different operating conditions and requirements, there are differences in formulation separating the two uses. Guns tend to operate at internal pressures of up to 450 MPa resulting in a very short burn time for the propellant. Rocket motors operate at 15 to 30 MPa and the propellant may burn for hundreds of seconds. For guns, high temperatures and residues are a major problem whereas this is less so for rockets.

Chemical propellants provide a simple and effective way of creating propulsion for flight. The earliest reports, from early in the millennium, are of the Chinese using rockets to confuse the enemy but the first true military use was by the Indians against British troops in the 18th century. By 1805 Sir William Congreve had devised a system for the British and these provided an important military advantage in the following decade. By the start of World War I, such rockets, all powered by gunpowder, had become obsolete. The return of the military rocket came during World War II from the desire to attack fast moving aircraft and to deliver strikes at great distances in greater concentrations than is possible from guns. Since then the importance of rocket powered weapons to attack on land, sea and in the air is unquestioned.

In this chapter the emphasis will be on battlefield weapons where no aspiration is required such as in ramjet engines. There are then two distinct categories of rocket propellants to consider, solid propellants and liquid propellants. The choice between these is based on the system requirements. Also, there is another distinction, this time in the end use of the energy. Chemical propellants are used both as the propulsive force to provide flight in a true rocket motor and as the energy source in power supplies. The power supply may well be within a missile or rocket taking the form of a small unit which provides power for the system to operate. Essentially, they provide hot gas under pressure which may execute a mechanical operation such as to spin-up a gyroscope for guidance or operate switches or pistons. There is a fine dividing line between power supplies and pyromechanisms. These will be considered separately in Chapter 8.

The performance of a rocket motor can be considered from several aspects. First, two parameters may be considered for the performance of the missile itself. One is the velocity at all burnt, the missile velocity when all the propellant is consumed and the other the maximum range. The latter of these depends on the type of flight, whether in an ideal ballistic trajectory or an airborne trajectory.

Another quoted value for the rocket motor is its thrust. This depends on specific impulse and on the mass rate of propellant consumption. The performance of a rocket motor, and therefore the missile, depends on the specific impulse (I_s) of the motor. This depends on the design of the motor and on the propellant used. Specific impulse is often quoted in either newton seconds per kilogram of propellant ($Ns\ kg^{-1}$) or simply metres per second ($m\ s^{-1}$).

Liquid Propellants

Types

There are two types of liquid propellant employed in rocket motors: they are monopropellants, in which a single component is present, and bipropellants, with separate fuel and oxidiser. These will now be considered separately.

Liquid Monopropellants

Liquid monopropellants consist of a single material which is capable of spontaneous decomposition with the evolution of heat and gas. The decomposition can be brought about either by simple thermal effects or in certain cases by the use of a catalyst. In general, these propellants have low specific impulse values: for example the I_s for hydrazine is 1950 $m\ s^{-1}$ and have found use only in such devices as attitude control thrusters on satellites or space probes. Currently the most favoured monopropellant is hydrazine (N_2H_4), a readily available chemical which can be effectively decomposed on a catalyst such as 30 per cent iridium supported on alumina. It can be stored for long periods in titanium alloy or stainless steel tanks and the only drawbacks to its use are its toxicity and a relatively high freezing point (2°C).

There are two decomposition reactions, the first (equation 6.1) is the exothermic decomposition of the hydrazine followed by an endothermic decomposition of a certain amount of the ammonia (equation 6.2). Under normal working conditions of a thruster motor, the catalyst will operate

$$3N_2H_4 \longrightarrow 4NH_3 + N_2 \tag{6.1}$$

$$2NH_3 \longrightarrow N_2 + 3H_2 \tag{6.2}$$

at about 1200°C and approximately 25 per cent of the ammonia will be further decomposed. Other monopropellants have been employed but have some drawbacks. Hydrogen peroxide was favoured before hydrazine became popular, however, it has storage decomposition problems. Nitromethane has better performance but is difficult to initiate and iso-propylnitrate (IPN) is prone to accidental ignition by adiabatic compression of trapped air bubbles when pumped.

Liquid Bipropellants

Liquid bipropellants consist of two components, a liquid fuel and a liquid oxidiser. They are injected into a combustion chamber where they are ignited either spontaneously by their own reaction (hypergolic) or by an igniter (non-hypergolic). They are very powerful systems and are used to power large rockets such as space vehicle launchers and in some tactical missiles such as the United States Army's Lance battlefield missile. The bipropellant system can be wholly or partly cryogenic, that is contain components which must be stored at very low temperatures of approximately -200°C. The other option is to utilise components which can be stored at ambient temperatures sometimes called earth-storable as opposed to the cryogenic space-storable.

Cryogenic systems are exemplified by the liquid hydrogen-liquid oxygen bipropellant. Their boiling points at atmospheric pressure are -252°C and -183°C respectively and are stored in special facilities to be filled into the rocket immediately prior to launch. The NASA Space Shuttle main engine and the third stage of the European Space Agency Ariane IV rocket employ this combination. It is not envisaged that this type of system would be used for battlefield weapons or even Intercontinental Ballistic Missiles (ICBMs).

Non-cryogenic bipropellant components can be sealed into containers for long periods and are described in this use as packaged liquid bipropellants. This allows constant readiness for use since the rocket is ready fuelled, an important requirement for a battlefield weapon. An example in current weapons is the system used to power the United States Army tactical missile known as Lance. This missile carries a warhead of approximately 500 kilogrammes and is powered by an unsymmetrical dimethylhydrazine (UDMH) - inhibited red fuming nitric acid (IRFNA) composition as fuel-oxidiser. UDMH is a derivative of hydrazine and IRFNA is a mixture of nitric acid, dinitrogen tetroxide with about 0.5 per cent hydrofluoric acid to inhibit corrosion of the storage tank. This hypergolic combination will give an I_s value of around 2700 m s^{-1} compared with a value of 3800 m s^{-1} for liquid hydrogen-liquid oxygen. Another example of a military use of a bipropellant system is the air-launched RB05A from the Saab Bofors Missile Corporation of Sweden.

Liquid rocket propellants have the advantage of high performance, clear exhaust, controllable thrust and, in some cases, cheap materials. However, there are also disadvantages such as complex plumbing and control and either cryogenic or toxicity problems. On balance, these propellants are best suited to power supplies or control motors using monopropellants and to very powerful rockets for space exploration, using bipropellants.

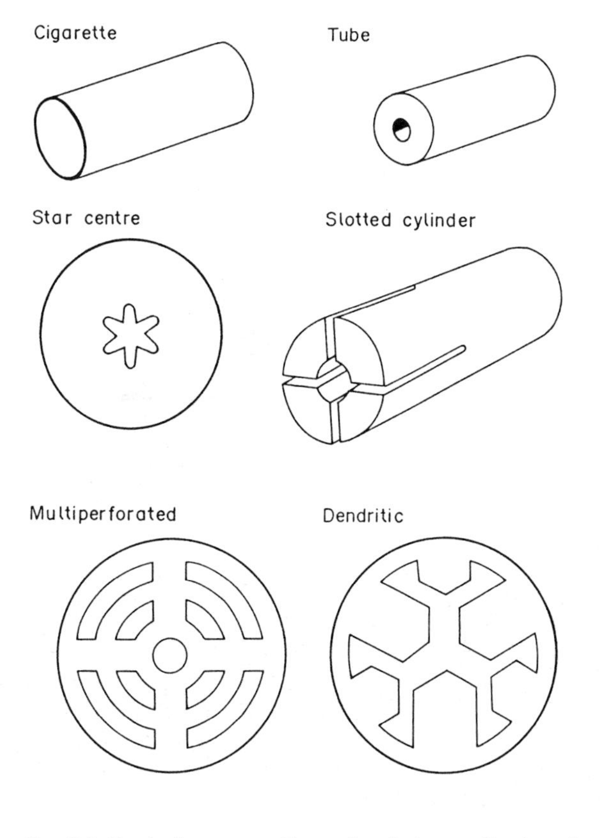

FIG. 6.1 Typical cross-sections of rocket propellant grains

Solid Propellants

Unlike liquid propellants where the propellant is pumped into a combustion chamber from storage, the solid propellant rocket motor must have all the propellant inside the combustion chamber on firing. Since we are considering a burning solid, the burning surface is important. Thus the control of a solid rocket motor is designed into the shape of the burning surface. It is usual for a solid propellant motor either to burn up a channel or channels in the filling or to end-burn like a cigarette. Thus, like gun propellants they are manufactured in specific geometrical shapes known, regardless of size, as grains. Some typical grain cross-sections and shapes are shown in Figure 6.1. To control the burning surface further it is essential that the flame cannot propagate around the outside of the grain and to prevent this one of two methods can be employed. One method is to bond the propellant grain to the motor casing (case-bonded) the other to have a flame-resistant sheath bonded to the grain (inhibited). Case-bonding is probably

FIG. 6.2 Short's ground-to-air Javelin missile

the preferred method although some classes of solid propellant are not amenable to this process. The chosen grain shape depends on the required motor performance. Motors can be considered from two viewpoints, boost and sustain. A boost motor needs to give rapid acceleration and have high thrust. The larger the burning surface the greater the mass flow rate of propellant: thus boost motors will have conduits to increase the initial and subsequent burning surface. In the extreme some motors need to burn before the missile leaves the launcher, sometimes called all-burnt on launch motors (ABOL). This is a requirement for shoulder launched systems to protect the soldier from the efflux. Depending on the weapon requirements there may then be a second stage boost motor which fires once the missile is clear of the operator such as in the Shorts ground-to-air Javelin missile shown in Figure 6.2. In other examples the ABOL motor is the only propellant and the missile then coasts to the target as for the United Kingdom light anti-tank weapon (LAW) shown in Figure 6.3. The sustain type of motor will contain either a solid grain which burns in cigarette fashion or, for higher performance, a star cross-section grain whose surface area remains essentially constant as the conduit burns outwards.

FIG. 6.3 United Kingdom Light Anti-tank Weapon (LAW)

The safe and efficient operation of a rocket motor depends on the motor design but also on the chemical and thermal stability, burning characteristics and mechanical strength of the propellant. Physical properties such as mechanical strength are just as important as chemical properties, cracking or shrinking may alter the burning surface. Thermal stresses, for example, are a particular problem as grain size increases. The requirements for solid rocket propellants are fulfilled by three families of materials as shown in Figure 6.4.

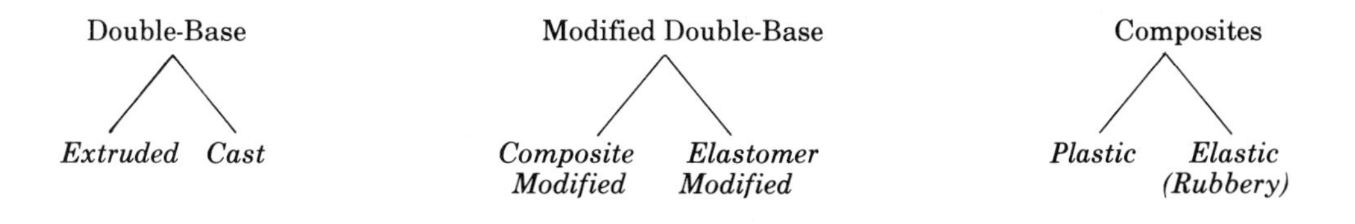

FIG. 6.4 Solid rocket propellants

Double-Base Propellants

THE NATURE OF DOUBLE-BASE PROPELLANTS

In most respects the double-base rocket propellants are like the double-base gun propellants discussed in Chapter 5. The explosive content is similar to the original cordite composition. Thus they are essentially a rigid colloidal mixture of nitrocellulose and nitroglycerine with a stabiliser such as ethyl centralite to remove oxides of nitrogen formed by slow decomposition in storage. Removal of these decomposition products is essential since they promote further decomposition (autocatalysis) and would give a very short storage life. Other additives include a plasticiser, such as diethylphthalate, to aid extrusion and prevent possible cracking, a burning performance modifier, like potassium sulphate, wax as a die lubricant and carbon black to prevent radiated heat from the burning surface causing premature sub-surface ignition.

MANUFACTURE

There are two methods of manufacturing grains of double-base rocket propellants, extrusion and casting. The extrusion method is similar to that employed in the manufacture of gun propellants and is limited to a maximum grain diameter of around 13 centimetres. Specialist large surface area grains for use in ABOL motors are also formed by extrusion. Table 6.1 provides a typical composition for an extruded double-base composition and this propellant would have an expected specific impulse (I_s) value of around 2300 m s^{-1}.

TABLE 6.1 COMPOSITION FOR A TYPICAL EXTRUDED DOUBLE-BASE ROCKET PROPELLANT

Nitrocellulose (13.25% N)	51.5%	Nitroglycerine	43%
Diethylphthalate	3%	Ethyl centralite	1%
Potassium Sulphate	1.25%	Carbon black	0.2%
Wax	0.05%		

The other method of manufacture, the cast process, relies on the use of a casting powder and casting liquid. The casting powder is similar to single base small arms propellant. It contains nitrocellulose, a stabiliser, such as 2-nitrodiphenylamine, and a plasticiser like dioctylphthalate. Unlike small arms propellant this powder may also contain a platonising agent; this will be considered later. The casting liquid is essentially nitroglycerine desensitised with glyceryltriacetate (triacetin). To form the cast double-base

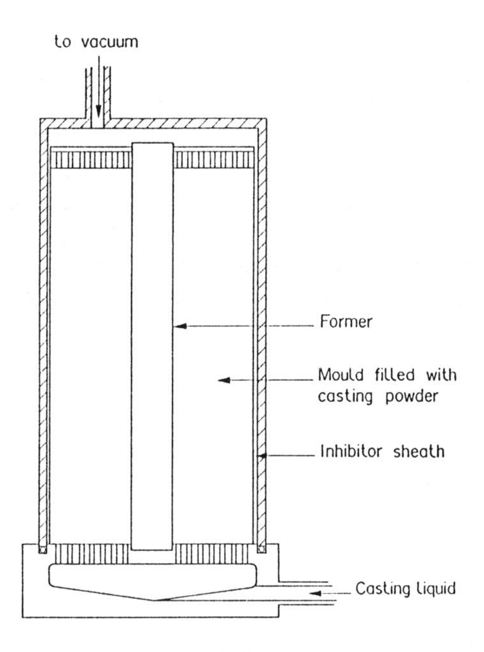

FIG. 6.5 Extruded rocket propellant

grain, a mould is filled with the powder and then the liquid is added under vacuum as shown in Figure 6.5. The powder absorbs the liquid and swells coalescing into a solid mass. Pressure is applied to give good consolidation and a curing process takes place at 60°C for a week or more. A composition for a cast double-base (CDB) is given in Table 6.2. These propellants have a higher percentage of nitrocellulose than the extruded type and will be of lower specific impulse (I_s) typically just in excess of 2000 m s^{-1}.

TABLE 6.2 COMPOSITION FOR A TYPICAL CAST DOUBLE-BASE (CDB) ROCKET PROPELLANT

Nitrocellulose (12.6% N)	59%	Nitroglycerine	24%
2-nitrodiphenylamine	2%	Triacetin	9%
Dioctylphthalate	3%	Lead stearate (platoniser)	3%

ADVANTAGES AND DISADVANTAGES

The double-base rocket propellants so far described have certain disadvantages. Low temperature cracking and gas bubble formation on storage are inherent problems. The extruded type cannot be case-bonded;

however, this is possible for most CDBs. Also, due to the large negative oxygen balance of this explosive composition, increasing the energetic output by adding aluminium alone is not possible. There are, on the other hand, distinct advantages to these compositions. The exhaust signature is minimal, with essentially smokeless burning. Complex grain shapes can be achieved with good rigidity. Also, due it is thought to the presence of nitrocellulose, additives can be included which modify the burning process such that the burning rate becomes almost independent of pressure over a given pressure range.

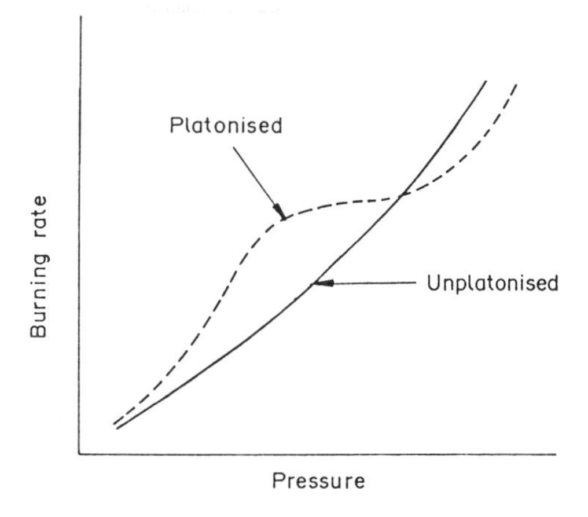

FIG. 6.6 Burning rate-pressure relationships

Figure 6.6 shows the burning rate-pressure relationships for unmodified double-base and for the modified or platonised composition. Platonisation is normally achieved by the addition of metal salts such as lead stearate and allows for less stringent control of the pressure within the motor.

Double-base rocket propellants are a popular choice for many systems. Extruded double-base propellant is used in the Shorts Brothers Javelin with an ABOL motor whereas the CDB type is used in the sustain motor for the French Exocet MM38 anti-ship missile and in the AJ168 British Martel.

MODIFIED DOUBLE-BASE PROPELLANTS

One of the problems with the standard double-base propellants is that the specific impulse performance is somewhat low for high performance missiles.

It is possible with explosives having an oxygen balance approaching zero to increase their power by the addition of aluminium powder. This is not possible for the double-base formulations; but, if another source of oxygen is added then that plus aluminium will improve significantly the motor performance. The oxidant of choice is ammonium perchlorate (AP) and when this plus aluminium is added to a cast-double base composition it is termed a composite modified cast double-base (CMCDB). When additives are made to CDBs they are sometimes said to be filled CDBs. The performance can be increased even further by addition of nitramine such as HMX. Table 6.3 gives a breakdown of two typical CMCDB formulations. These will have specific impulse values of over 2600 m s^{-1}, however, there are penalties incurred by such modifications. These propellants can no longer be platonised and have a smoky efflux. The smoke arises from the presence of aluminium oxidising to solid aluminium oxide together with hydrogen chloride, which is a consequence of the ammonium perchlorate.

TABLE 6.3 Compositions for Typical Composite modified Cast Double-base Propellants

	Per cent	*Per cent*
Nitrocellulose (12.6% N)	20	22
Nitroglycerine	30	30
Triacetin	6	5
Ammonium perchlorate	11	20
Aluminium	20	21
HMX	11	
Stabiliser	2	2

Hybrids

The CMCDBs are examples of hybrid rocket propellants. Another hybrid type has been developed by IMI Summerfield Ltd in the United Kingdom. Again, it is a modification of the conventional CDB but in this case the object is to improve the physical properties. One of the drawbacks of CDBs is that they are prone to low temperature cracking. The addition of an elastomeric polymer to the colloidal matrix gives low temperature strain capabilities and thus overcomes this problem. The extensibility of an unmodified CDB at –50°C will be less than 3 per cent whereas that for an EMCDB will be around 30 per cent. Environmental tests on production motors will be performed in the range –50 to +70°C. These elastomer modified cast double-base (EMCDB) propellants lose none of the advantages of CDBs such as platonisation and case-bonding. The EMCDBs can also be filled to improve performance as in CMCDBs without losing their low temperature elasticity.

Composite Propellants

Polymer Addition

The double-base propellants have, as their energy source compounds which are explosives in their own right: that is, nitrocellulose and nitroglycerine. However, after World War II, it was realised that the cordite type of compositions were not ideal rocket propellants and that alternatives should be sought. Since propellants do not detonate in their correct role, it was considered that the fuel-oxidiser intimacy found in explosive molecules might not be necessary and that fuel-oxidiser mixtures could give as much energy. It was also realised that physical properties for this type of propellant were critical and thus the newly discovered polymers might provide a suitable fuel. Two classes of polymer are available: those which are solid at ambient temperature but will flow at higher temperatures and those which are cross-linked on final polymerisation and do not flow on heating. When they are used in propellants, the first type are known as plastic and the latter rubbery or sometimes elastic materials. The oxidiser is then chosen on the grounds of thermodynamic requirements, physical properties and safety.

The polymer or binder in a composite propellant usually accounts for only 10-15 per cent by weight of the composition. However, this binder largely determines the mechanical properties of the final material. Perhaps the most important of these properties is the elastomeric behaviour since it is likely that the case of the motor and the propellant expand and contract at different rates during heating and cooling cycles. If the propellant cannot take up this difference in behaviour, it may well detach from the motor and case-bonding will be lost. This would allow the flame front to burn over surfaces previously unavailable and the motor might well over-pressurise and explode. This may also be the consequence if cracking of the propellant occurred. This could come about by the inability of the propellant grain to absorb internal stresses. Thus, the general requirements for the binder are high tensile and compressive strength with good elasticity. It must also be capable of accepting a high solids loading and be amenable to manufacturing processes. The material to be used in manufacture should be, ideally, a liquid at relatively low temperatures (~100°C) with a workable viscosity (1-10 Ns m^{-2}) which converts to a solid on either cooling or curing. It also must be non-hygroscopic, be compatible with other ingredients and produce low molecular weight species with high heats of formation.

Oxidisers

The oxidiser makes up the largest part of the composite propellant for up to 85 per cent by weight in some cases. Of the solid oxidisers available, those with metal cations such as sodium nitrate, potassium perchlorate etc are unsuitable since the metals will take up oxygen giving solid products and less heat output than if the oxygen was used to oxidise carbon or hydrogen. Early formulations used ammonium nitrate but it soon became clear that the best candidate for high energy output was ammonium perchlorate. This salt is essentially non-hygroscopic, has a high solid density (1.9 g cm^{-3}) and, on combustion, liberates not only oxygen required to oxidise the fuel matrix (equation 6.3), but also 193 kJ $mole^{-1}$ of heat energy.

$$2NH_4ClO_4 \longrightarrow N_2 + Cl_2 + 4H_2O + 2O_2 \qquad (6.3)$$

Other contenders such as chlorates are unsuitable since mixed with a fuel they are too sensitive to use. Ammonium chlorate as such does not exist.

The ability of aluminium to increase the specific impulse of modified CDBs when ammonium perchlorate is also added can now be seen to be a true hybrid of these composite propellants with the double-base variety. Thus composites can have aluminium added to great advantage since it is only another fuel to be chemically balanced together with the binder against the oxidising ability of the ammonium perchlorate.

Plastic composites

The first generation composite propellants were of the plastic type and thermoplastic polymers such as polystyrene and polyvinylchloride were employed, followed by polyisobutene (polyisobutylene) (PIB). The latter material became the polymer of choice which, to aid manufacture, has a plasticiser added such as ethyl oleate or dioctyl sebacate, giving a mixture often termed plastisol. The balanced formulation for maximum output should contain ~ 10 per cent binder/ ~ 90 per cent oxidiser ratio for plastisol/ammonium perchlorate. In fact, the binder is present at about 15 per cent by weight due to an inadequate 'solids capacity' of the binder. However, this type of propellant is very simple to manufacture and there is no potential hazard of exothermic reactions occurring within the polymer, since no curing reaction is necessary. The ingredients are mixed in an incorporator at 70°C for around one hour then extruded into a vacuum chamber to de-aerate the mixture. The plastic propellant is then loaded into a hydraulic press and extruded into the motor case, which is pre-coated with adhesive to give the case-bonding. This is a simple process compared with the rubbery composites developed later. However, these newer composites have superior mechanical properties, particularly as motor size increases. Plastic composites can slump under their own weight and, in practice, these materials are used in motors of less than 15 centimetres diameter.

Other additives which may be present in plastic composites include aluminium and burning rate modifiers. Often the composition will burn too quickly or at too high a temperature. To cool the burning, ammonium picrate is added, often in high percentages. This also has the effect of reducing the specific impulse. Copper chromate is another additive; this time it is to catalyse the burning process and thus increase the burning rate. Oxamide may be added to produce a slight platonising effect although only double-base propellants can be truly platonised. Finally lecithin is added to act as a surfactant ensuring good contact between ingredients. Table 6.4 gives some typical composite propellant compositions.

TABLE 6.4 TYPICAL COMPOSITIONS OF COMPOSITE PROPELLANTS

	Per cent	*Per cent*
Ammonium perchlorate	42	57
Ammonium picrate	38	30
Plasticised PIB	12	12
Aluminium	5	—
Oxamide	2	—
Lecithin	1	1

The advantages of plastic composites are that they are cheap to produce and can be case-bonded. They have a high specific impulse compared to DB propellants and can be aluminised to give specific impulse values of ~ 2600 m s^{-1}. A good shelf life is also expected. However, they have poor stress resistance at low temperatures, particularly compared with rubbery composites and are unsuitable for large motors due to slumping. Like all composites they create a smoky efflux and they cannot be platonised. On balance the plastic composite is less favoured than other alternatives and is not often used. One example of an in-service use is in the United Kingdom Bloodhound missile.

Rubbery Composites

The rubbery composites work on a very similar principle of a hydrocarbon polymer fuel with ammonium perchlorate as oxidiser. In this case, however, the polymer becomes cross-linked in a curing stage and thus has unique physical properties. The development of these propellants owes much to research in the field of polymer chemistry, which provided materials with the physical properties desirable for propellant grain production. The first rubbery type used polyurethanes (PU) as the fuel matrix and the method of preparation will serve to illustrate the general method of rubbery composite propellant production. The starting material is already a polymer, in fact a polyester, which has a molecular weight of about 2000 and is a viscous liquid at room temperature. The viscosity is such that the solid propellant components can be readily incorporated, that is the ammonium perchlorate oxidiser plus any additives such as aluminium powder, plasticiser, nitramine etc. After warming and de-gassing under vacuum, this slurry is then cooled and a cross-linking agent added; in this case it is a diisocyanate. The composition is then cast under vacuum into a motor case with formers to give the internal channels as required. The mould and contents are then held at an elevated temperature, ~ 60°C, for several weeks. During this time the following reactions shown in equations 6.4 and 6.5 occur. The diisocyanate reacts with two alcohol groups on the ester molecules. This may link two polyester chains end to end increasing the molecular weight.

$$\text{HO}\frown\text{OH} + \text{OCN-R-NCO} + \text{HO}\frown\text{OH} \rightarrow \text{HO}\frown\underset{\text{O}}{\text{OC}}\text{-}\underset{\text{H}}{\text{N}}\text{-R-}\underset{\text{H}}{\text{N}}\text{-}\underset{\text{O}}{\text{CO}}\frown\text{OH} \qquad (6.4)$$

This in itself will form a rubber but without the correct elasticity. The reaction of hydroxyl groups positioned along the polyester chain gives cross-linking to result in a three dimensional structure and a matrix structure which gives the required physical properties. The control of the extent of cross-linking is important since highly linked structures will be brittle. It should be noted that the cross-linking reaction of isocyanate with the hydroxyl group gives no other product.

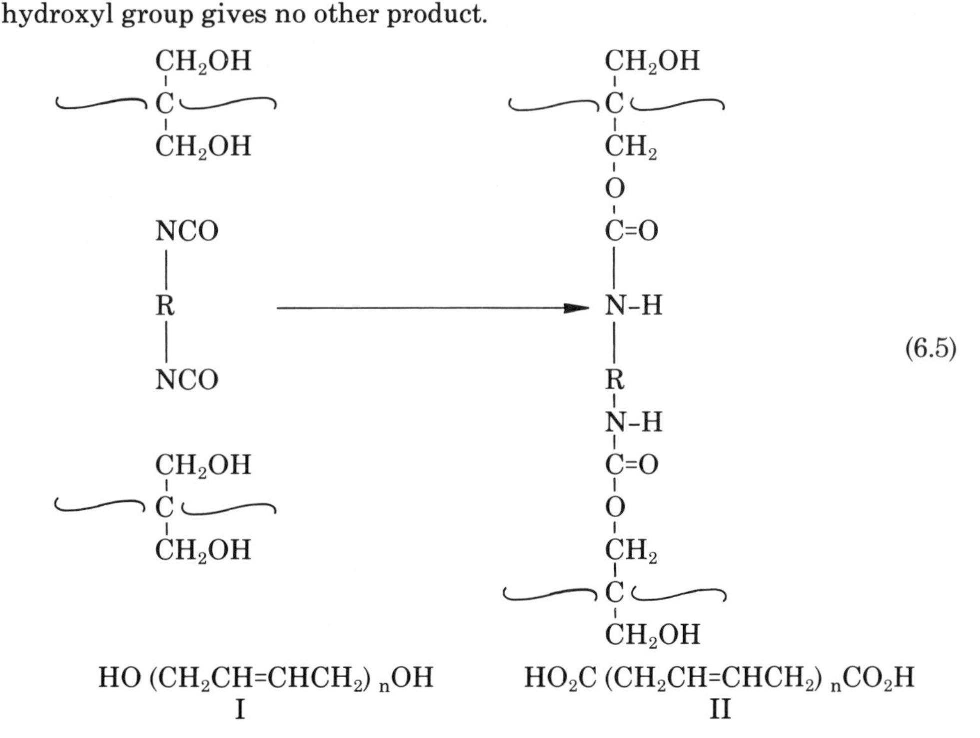

$$\text{HO}\,(\text{CH}_2\text{CH=CHCH}_2)_n\text{OH} \quad \text{I} \qquad \text{HO}_2\text{C}\,(\text{CH}_2\text{CH=CHCH}_2)_n\text{CO}_2\text{H} \quad \text{II}$$

Polymer Composites

This first generation rubber composite has now largely been overtaken by types based on polybutadiene. There are two sub-groups; hydroxyl terminated polybutadiene (I) (HTPB) and carboxyl terminated polybutadiene (II) (CTPB) based polymers.

The polyurethanes are pre-polymers and, like polyester, they are liquids into which the solids can be mixed before cross-linking. The cross-linking for HTPB polymer agent, is a diisocyanate and the chemistry is similar to that described above. Other cross-linking agents, such as epoxides (III) or aziridines (IV), must be employed to cross-link the CTPB pre-polymer. These will link two pre-polymer molecules via their carboxyl end-groups and, if cross-linking agents are themselves polyfunctional, will cause true cross-linking, giving the rubber matrix. Another carboxyl terminated pre-polymer

```
        O                      R'
       / \                     |
   CH2-CH-R                    N
                              / \
                          CH2-CH-R
      III                    (IV)
```

finding favour is a poly(butadiene-acrylic acid-acrylonitrile) material known as PBAN. This will be cross-linked by epoxides or aziridines. Again, the curing process takes place after the mixture is cast into the motor body and takes several weeks at 60°C. Table 6.5 gives some data on typical rubbery composites.

The HTPB and CTPB propellants have excellent mechanical properties over a wide temperature range (–55 to +70°C) and case bond extremely well. For example, extensibilities of up to 35 per cent at the lower end of this temperature range are claimed. It is more practical to form propellant grains by casting, although by careful control of conditions it is possible to extrude some of these materials before curing. They can be effectively aluminised and

TABLE 6.5 Composition (percent) and Performance of Some Typical Rubbery Composite Propellants

Components	*PU*	*PBAN*	*CTPB*	*HTPB*
Ammonium perchlorate	70	69	63	70
Matrix polymer	21	11	10	12
Aluminium	8	15	17	18
Di-octyladipate		4		
HMX			10	
Bonding agent	1	1		
Specific impulse ($m\ s^{-1}$)	2400	2260	2600	2550

the rubber matrix has a good solids capacity. Smooth burning, particularly when aluminised, is another advantage. There are disadvantages to be considered. The curing process must be controlled very carefully if the optimum physical properties are to be achieved and stress cracking on storage has been noted. One of the major drawbacks is the very smoky exhaust from all composite rocket motors due to the likelihood of hydrogen chloride formation in the products as the motor burns. This gas will immediately form droplets with water forming an aerosol or smoke giving a large white plume. If aluminium is present the solid product, aluminium oxide, will increase further the smoke output. The exhaust plume also may interfere with electromagnetic radiation; thus guidance of rockets powered by composite propellant motors from the firing point may be a problem. The interference is caused by ionic species present at the high temperatures found at the rocket nozzle. These species are probably ionised atoms or molecules but may also include free electrons.

The list of applications for rubbery composites is enormous and thus a representative sample is mentioned. Examples of the use of HTPB types are in the American high-speed anti-radiation missile (HARM) AGM-88A and in the large surface area grain ABOL motor of the British LAW 80. The latter is an example of an extruded rubbery propellant. The French Super 530 air-to-air missile uses a CTPB propellant and the American Shrike, AGM-35 air-to-surface missile uses a PU propellant in the improved MK 78.

7. Pyrotechnics

Beginnings

The dictionary definition of pyrotechnics is 'the art of making and using fireworks'. It is not surprising, therefore, that military pyrotechnics are considered by many to be large versions of the devices which we call fireworks. There may be similarities in the types of output. However, there the similarity ends. To produce fillings for military pyrotechnics very high energy compositions are used, making these munitions truly explosive in nature.

The pyrotechnic explosive compositions date back to antiquity and gunpowder. The first recorded explosive, by any definition is a pyrotechnic. Fireworks for pleasure developed over the centuries particularly in the East and, by the 18th century, large displays were common in Europe. However, it was by the mid-19th century that the use of pyrotechnics for signalling began. Since that time, the military use of pyrotechnics has increased and been refined.

Meanings

A literal translation of pyrotechnic could be heat or explosive actuated. This, of course, is far too general, but it should be noted that in some texts pyromechanisms containing detonating high explosives are called pyrotechnics. A clear distinction between pyrotechnics and pyromechanisms is difficult. However, consideration of the definition below together with the content of Chapter 8 on pyromechanisms should be sufficient. Most pyrotechnics can be considered as '*materials capable of combustion when correctly initiated to produce a special effect*'. There are three points to note from this definition. First, these compositions should burn and not detonate, but second, it is important to consider initiation. Some compositions are capable of detonation if heavily confined or incorrectly initiated by over-stimulus. The third point is what distinguishes pyrotechnics from other explosive types and that is they provide special effects.

Effects

The special effects provided by pyrotechnics can be categorised as shown in Table 7.1. All will produce heat, although for some uses the heat output is the effect to be utilised. Light output could be regarded partly as a heat effect since high temperatures are required to create the radiant energy; also screening and signalling smokes are heat dependent. The category labelled 'light' should in reality be 'electromagnetic radiation' since wavelengths other than those in the visible spectrum are relevant. Infra-red radiation is of great importance in the production of decoys.

TABLE 7.1 PYROTECHNIC SPECIAL EFFECTS

Effect	*Examples*
Heat	Igniters, Incendiaries, Delays, Metal producers, Heaters
Light	Illumination (long and short periods), Tracking, Signalling, Decoys
Smoke	Signalling, Screening
Sound	Signalling, Distraction

Compositions

Basic Components

In most cases pyrotechnic compositions consist of a fuel and an oxidiser, together with other additives to give the special effect as required. Nitrated organic materials, such as tetranitrocarbazole (TNC) and nitroguanidine, occur in some compositions but this is relatively rare. Also, there are materials which do not burn and yet are considered as pyrotechnics. This is because they produce an effect also created by the true pyrotechnic compositions. For example, titanium tetrachloride ($TiCl_4$) reacts with moisture in the air to form a screening smoke. It is a hydrolysis and not a combustion. Red and white phosphorus burn in air to give a screening smoke and thus is considered a pyrotechnic event. It is an example of a 'fuel only' composition.

Various Combinations

Unlike high explosives or propellants, where only a small number of different materials are used, for pyrotechnics the range is extremely large allowing careful tailoring of the composition. The fuels are usually powdered elements, either metals or non-metals, all of which when oxidised provide heat energy. A selection of fuels is given in Table 7.2. The oxidisers are also varied and can be considered as classes of salts, examples are given for each type in Table 7.2. Most are sources of oxygen. However, halocarbons can also act as oxidising agents in the true chemical sense, the halogen fluorine is the most energetic chemical oxidiser. As part of the formulation there are other components which can be just as important to the final composition as the fuel-oxidiser system.

TABLE 7.2 FUELS AND OXIDISERS USED IN PYROTECHNIC COMPOSITIONS

Fuels		*Oxidisers*	
Metals:			
Aluminium	(Al)	Chlorates	($KClO_3$)
Chromium	(Cr)	Chromates	($BaCrO_4$)
Iron	(Fe)	Dichromates	($K_2Cr_2O_7$)
Magnesium	(Mg)	Halocarbons	(C_2Cl_6, PTFE)
Manganese	(Mn)	Iodates	($AgIO_3$)
Titanium	(Ti)	Nitrates	(KNO_3)
Tungsten	(W)	Oxides	(BaO_2, ZnO)
Zirconium	(Zr)	Perchlorates	($KClO_4$)
Non-metals:			
Boron	(B)		
Carbon	(C)		
Silicon	(Si)		
Sulphur	(S)		
Phosphorus	(P)		

Binders

Binders do more than just consolidate a composition. The technology and the earlier art of pyrotechnics relied on a knowledge of binder materials. Table 7.3 lists binders showing the diversity of choice. Some of the waxes used for centuries are still popular and are natural in origin. More recently there has been a move to man-made polymers. These resins and rubbers can be more closely controlled for quality and manufacturers are not dependent on overseas suppliers. As to their role, binders do increase cohesion between particles aiding consolidation. Nevertheless, an equally important function is to coat and protect reactive components, such as the metal powders, which may react with oxygen or moisture. They also modify the burning rate and so the performance and reduce sensitiveness to stimuli, such as friction. In some cases, they can enhance performance, such as in illuminating flares.

TABLE 7.3 BINDERS USED IN PYROTECHNICS

Natural	*Man-made*
Paraffin wax	Bakelite resin
Beeswax	Polyester resin
Carnauba wax	Chlorinated rubber
Chinese wax	Polyvinylchloride
Boiled linseed oil	Thiokol rubber
Lithographic varnish	Epoxy resin
Shellac	

In the manufacture of any consolidated pyrotechnic composition, which is known as a candle, the binder may simply be added into the mix. It is also common first to coat a constituent, usually the metal powder, by evaporating an organic solution of the binder on to the powder which, after drying is mixed with the other ingredients. The man-made binders are often cured after mixing to give a cross-linked rigid polymeric structure. This may give large temperature rises on polymerisation. Even the natural materials can create heat when used. Both lithographic varnish and boiled linseed oil oxidise in air with evolution of heat. This reaction must be allowed to finish before consolidation.

Material Properties

Problem of Degradation

In general, pyrotechnic compositions are far more susceptible to degradation in storage than other explosive materials. This is not surprising, since about two-thirds of all formulations contain finely divided metals, often with high reactivity, such as magnesium. These metals have the property of reacting with moisture, particularly in salt solutions.

Salts

Most oxidisers are salts of some variety. To compound the problem, these oxidiser salts may well be hygroscopic and thus attract moisture from the atmosphere. This means that finished stores must be sealed. However, if moisture is already present, for instance in cardboard liners, then the reaction of the metal with water produces hydrogen gas. This may pressurise the device causing either a dangerous pressure build-up, or perhaps breaking the integrity of hermetic sealing allowing more atmospheric moisture to enter. All these things may render the device unreliable. The complete reaction of one-tenth of a gramme of water, one drop, can produce 60 millilitres of hydrogen gas. This emphasises the importance of lacquers or varnishes used to coat and protect metal powders.

Impurities

Other less expected reactions may occur with pyrotechnic components. For example, some screening smokes use zinc oxide (ZnO) in their formulation. It can react with carbon dioxide from the atmosphere forming zinc carbonate ($ZnCO_3$) which can drastically modify the burning characteristics of the smoke mix. Such an impurity may well be present in the stored material prior to formulation and thus quality control of ingredients is vital in the preparation of pyrotechnics. It is not uncommon for a batch of pyrotechnic devices to have their performance outside the specification range due to the presence of impurities or to reactions on storage.

Chemical Reactions

The possibilities for the occurrence of chemical reactions are manifold and the end result may not be just a degradation of performance, it may be a spontaneous ignition. It is not usually recommended to use either sulphur or phosphorus in combination with a chlorate salt as oxidiser, because both fuels can form acids in the presence of moisture and oxygen. The acid then reacts with the chlorate to form the highly reactive and unstable chloric acid. Compositions do exist with chlorate and sulphur but always containing an acid neutraliser such as calcium carbonate. A judicious consideration of any possible chemical reactions should indicate which compositions are the safest.

Sensitiveness

When considering safety of an explosive material, it is necessary to consider the sensitiveness to the variety of stimuli already covered in Chapter 3. In general, pyrotechnic compositions are mixtures of highly reactive chemical compounds or elements. The particle sizes are usually very small, microns or tens of microns, and thus have a large surface area for reaction. They are also often sharp or gritty, being salts or metal powders. It is not surprising, therefore, that these compositions are often quite sensitive to various types of mechanical energy input.

As a generalisation, pyrotechnics are most sensitive to friction, flame and electrical discharge and perhaps to a somewhat lesser extent, percussion. It is probable with these materials that there is a significant frictional component in the percussion initiation. The presence of certain ingredients will often point to sensitive compositions such as potassium chlorate. Other materials are well known for sensitiveness to particular stimuli. Zirconium containing compositions are renowned for their spark sensitiveness and full static precautions must be taken when they are manufactured and filled into stores.

Thus sensitiveness is dependent on which materials are used in the formulation. However, even when the type of materials has been chosen, other factors also are important. For a given composition, changes in the particle size of one or more ingredients may alter sensitiveness as may the intimacy of mixing. Alternatively, keeping the properties of the various ingredients constant but altering the mixing ratio may also change susceptibility to initiation. These changes may alter sensitiveness but they will have a significant effect on the performance of the pyrotechnic.

Factors Affecting Performance

Constituents

Obviously, the performance of any pyrotechnic composition will depend on the chosen constituents. What is under consideration here are the factors which can influence the performance after the basic ingredients have been chosen. Since a burning pyrotechnic composition is in fact a chemical reaction occurring between the components, the first factor to consider is the ratio or proportions of these reactants.

It can be said that any chemical reaction can be balanced by choosing the correct ratio of reactants, so that all the reactants are consumed to give the products. The ideal or correct ratio of reactants is called the stoichiometric mix. At, or very near, this ratio, the composition will give its maximum performance. Since these reactions are exothermic, it will also equate to the maximum heat output which in turn will give the fastest burning rate. Any composition change away from stoichiometry, either to fuel or oxidiser rich, will reduce the heat output from the reaction. This will affect the overall performance of the composition.

In some cases this concept applies quite straightforwardly such as for the Si-PbO_2 system where the silicon percentage for stoichiometry is around 9 per cent and maximum heat output is at about 10 per cent Si. However, if oxygen in the atmosphere can be considered a reactant, such as in Mg-$NaNO_3$ flares, then the maximum output may well occur at magnesium percentages far higher than expected. Also when attempting to calculate the optimum mixture it may well be difficult to ascertain the chemistry occurring. This is because chemistry at over 2000°C is not the type usually considered by chemists. The most stable products at more normal temperatures may well not form at these extremes. There have been cases where the stoichiometric ratio has been found experimentally; the composition has been found to give maximum heat output, and the chemistry rationalised from this.

Burning Rate

The performance of any composition of most importance to the pyrotechnician is the burning rate. The maximum burning rate is often slightly to the fuel rich side of the stoichiometric mix, particularly if the fuel is a metal powder. This is due to enhanced thermal transfer allowing the burning layer more readily to heat the next layer and bring it to ignition temperature. The shift is greatest for metals, then metalloids and negligible if the fuel is a non-metal such as carbon or sulphur.

Since the burning rate has a maximum value for a given percentage of fuel then if the fuel content is either increased or decreased, burning rate will be slower. Consequently two compositions will have the same burning rate, as shown in Figure 7.1, and thus a choice must be made. Logically it would be preferable to choose the one with the higher percentage of fuel since it would probably be easier to mix and obtain homogeneity. Also on the fuel rich side of the curve the slope is less steep and thus slight changes in composition caused by poor mixing would have less effect on the burning rate.

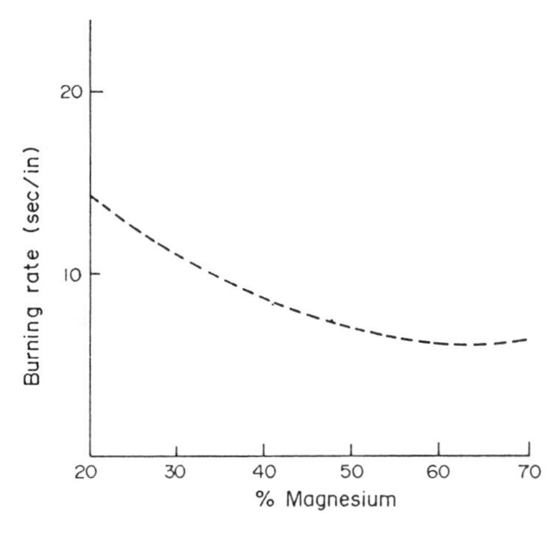

FIG.. 7.1 Burning rate versus stoichiometry

Retardants

The composition in which small composition changes should have least effect is at the top of the curve with the stoichiometric mix. However, this may well burn too quickly for the desired performance. One way that this performance may be altered without changing stoichiometry of the reactants is to add an inert material. This is common practice and it is usual to add materials such as clays, for example kaolin to slow burning rates.

Another type of additive used to slow compositions decomposes endothermically reducing the overall heat output. A popular example of this type of retardant is calcium oxalate which decomposes to calcium oxide, carbon monoxide and carbon dioxide. It is usually added as the monohydrate further adding to the cooling effect; see equation 7.1.

$$Ca(C_2O_4) \cdot H_2O \longrightarrow CaO + CO_2 + CO + H_2O \qquad (7.1)$$

Any material added to a composition not directly involved with the fuel-oxidiser reaction will modify the burning rate. This will include binders and it has been noted that the addition of different binders, in the same percentage, to some compositions, gives very different burning rates.

Particle Size

Of great importance when considering materials is the specific surface. In the simplest of terms this is particle size since we must consider the surface area of the materials. It is easier for a reaction to take place on a large surface area. Thus, the smaller the particle size of the ingredients the more rapid the reaction. Metal powders are available from sub-micron sizes, often called ultrafine, upwards. The usual range for metal powders is between 5 μ and 1000 μ and these are produced by blowing compressed gas through the molten metal giving it the name of blown or atomised material. Particles are often approximately spherical although flake forms exist. However, simple surface area criteria do not fully explain material reactivity. Firework manufacturers often use a grade of aluminium called dark pyro because of its good reaction properties: it will often out-perform blown aluminium of smaller particle size.

What determines the reactive surface, or sites, is far more complex than simple variations of surface area to volume of the macro-world. Reactivity occurs at the atomic or molecular level. For solids to react as is often required in pyrotechnic chemistry, atoms or molecules must be pulled away from the surface, and those with the least attractive forces holding them back will be most available. There is also an order of bond strengths to be considered for the different types of materials. The strongest bonding in solids is found in covalent lattices such as silicon. Ionic lattices come next covering metal oxides and salts. This is followed by metallic bonding in elemental metals and finally the weakest bonding is found in crystalline or amorphous solids made up from discrete molecules such as sucrose or hexachloroethane.

Micro-structure

Reactivity is highly dependent on the intimate structure of the solid components. Any defects within the lattice structuring is a weakness and may enhance reactivity. These defects may arise naturally as in non-stoichiometric compounds such as zinc oxide, or may be caused by impurities. Impurities may be present by accident, or can be engineered into the compound by doping. For example, germanium reacts with molybdenum trioxide as shown in equation 7.2. If the germanium is doped with small amounts of arsenic, the reactivity is markedly increased.

$$Ge + 2MoO_3 \longrightarrow GeO_2 + 2MnO_2 \qquad (7.2)$$

Another example is the reaction of potassium chlorate doped with copper chlorate when mixed with sulphur. After mixing, a spontaneous reaction occurs on standing; whereas the un-doped potassium chlorate when mixed with sulphur, will store for long periods without reaction. It should be noted here that this mixture is extremely sensitive and it is not advisable to use this combination.

Heat and its Applications

Heat of Combustion

All pyrotechnics produce heat as they burn. In many uses or special effects, this production of heat is the actual requirement. Since the majority of compositions contain a metal as fuel, a first consideration of heat output is the heat of combustion of the metals. Table 7.4 gives a representative sample of heats of combustion of relevant metals and non-metals. Thus some elements are prodigious energy producers whilst others have a modest value. Note also that density plays a part if fixed volumes are considered.

TABLE 7.4 Heats of Reactions of Metals, Metalloids and Non-Metals with Oxygen (Heat of Combustion), Sulphur and Selenium kJ g^{-1} (kJ cm^{-3})

	Oxygen Product		*Sulphur Product*		*Selenium Product*	
Metals						
Al	Al_2O_3	31.3 (84.5)	Al_2S_3	9.4 (25.4)		
Mg	MgO	24.8 (42.2)	MgS	14.3 (24.3)	MgSe	2.5 (4.25)
Ti	TiO_2	20.0 (89.8)				
Zr	ZrO_2	12.1 (77.2)				
Fe	Fe_3O_4	7.5 (58.9)	FeS	1.7 (13.4)		
W	WO_3	4.6 (89.2)	WS_2	1.1 (20.3)		
Metalloid						
B	B_2O_3	57.8 (70.8)				
Si	SiO_2	31.3 (62.5)				
Non-metal						
C	CO_2	32.7 (65.3)				
P	P_4O_{10}	24.0 (43.7)				
S	SO_2	9.3 (18.6)				

On a volume basis, tungsten (W) is as energetic as aluminium (Al), whereas on a weight basis tungsten is much the inferior of the two. The largest energy outputs for compositions are derived from mixtures of the elements, usually metals, with oxidisers of the salt type. These oxidisers are often metal salts of nitrates, chlorates or perchlorates and are high in oxygen content and low in ballast, elements not taking part in energy producing reactions. Examples are counter-ions (Na^+, K^+ etc), nitrogen or chlorine, all with low atomic weights.

Also the energy required to decompose the oxidisers is either small for nitrates, approximately zero for perchlorates and for chlorates heat is actually evolved. Commonly heat outputs of over 8 kJ g^{-1} are observed for these mixtures. For a given fuel, the chlorate oxidisers give the greatest output, however, compositions containing chlorates are usually more sensitive to initiation than those containing nitrate or perchlorate.

Thermitic Reactions

Another important group of oxidisers are the metal oxides. They will often react with metals the so-called 'thermitic reactions'. An example of such a reaction is shown in equation 7.3.

$$2Al + Fe_2O_3 \longrightarrow Al_2O_3 + 2Fe \tag{7.3}$$

From a thermodynamics viewpoint, the energy output of this type of reaction depends solely on the heats of formation of the two oxides. Thus heat of reaction (ΔH_r) is the difference between the heat of formation of Al_2O_3 and that for Fe_2O_3, (ΔH_f Al_2O_3 - ΔH_f Fe_2O_3). The greater the heat output, the more vigorous the reaction. As a general rule, a steady reaction will occur if the difference between the heats of formation of the two oxides is about 120 kJ mole^{-1}. If the difference is greater than 160 kJ mole^{-1} the reaction tends to be violent as in the reaction of aluminium with lead dioxide. On the other hand, as the difference gets smaller, the reaction becomes sluggish and probably will not go to completion. Table 7.5 gives some theoretical outputs for metal/metal oxide combinations.

TABLE 7.5 THEORETICAL HEAT OUTPUTS FOR SOME METAL/METAL OXIDE COMBINATIONS, kJ g^{-1} (kJ cm^{-3})

Metal	*PbO_2*	*CuO*	*MoO_3*
Al	3.1 (21.3)	4.1 (21.0)	4.7 (18.4)
Ti	2.4 (18.1)	3.1 (17.8)	3.1 (14.4)
Mg	3.9 (16.8)	4.3 (16.5)	4.8 (14.2)
Mn	1.1 (2.3)	1.4 (8.8)	0.7 (3.9)
W	0.8 (8.6)	0.9 (1.9)	0.2 (0.5)

Although it is not feasible to mix metals with elemental oxygen a reaction also occurs between metal powders and the chalcogens (sulphur, selenium and tellurium). These reactions have a modest heat output in general although the best oxidiser (sulphur) together with metals giving large heat outputs when reacting with oxygen (see Table 7.4) such as aluminium give quite respectable reactions. Table 7.4 also shows the types of output possible for some of these combinations. The organic halogens are another group of oxidisers used in reactions with metal fuels. For example one of the first smoke producing compositions utilised the reaction between zinc powder and carbon tetrachloride, the so-called Berger mixture. More recently magnesium and polytetrafluoroethylene (PTFE) have provided a prodigious energy producer used in decoy flares.

First Fires

Requirements

Pyrotechnic devices generally consist of the main composition pressed into a container. On top of this is added a layer of composition; it is designed to accept the output of an initiator such as an electric fuze head. This layer has many names: first fire, igniter, priming composition, and starter mixture are examples. Its role is to act as a booster between the initial low energy output of the initiator and the main composition which may be difficult to ignite. The electric fuze head will probably initiate most loose powder compositions; however, the pressed candles usually require a first fire.

The requirements for the first fire are surprisingly quite stringent. It must be easy to ignite by the initiator output and generate a large amount of heat but not too rapidly or violently. A violent reaction may cause the first fire to blow off the surface which is to be ignited and cause a failure. It is advantageous to produce some non-volatile products to leave a hot slag and thus transfer the heat to the main composition. The components must be compatible with those in the main charge since the first fire will be either pressed as a final layer, or laid on to the surface of the main charge. For example, a first fire containing sulphur should not be used with a main charge containing chlorate. Finally, the composition should not be too sensitive and should have an output matched to the relative ease of initiation of the main filling. This is achieved by three types of composition.

Gunpowder and its Variants

The first of these is gunpowder and its variations. They will pick up flash very easily and have a suitable output, which produces hot gases and thus pressure and also hot particles. A popular method of employing gunpowder is as an impregnated material known as Cambric. An enlarged view of this material is shown in Figure 7.2.

Sulphur-free gunpowder is also produced for first fires where the main charge would be incompatible with sulphur. This is particularly important for compositions containing magnesium as this reacts on prolonged contact with sulphur.

Organic Compounds

A second group, which are generally much cooler than gunpowder, is organic compound plus oxidant. A good example of this is a 60/40 composition of potassium nitrate and lactose, used to ignite coloured smoke compositions where over-ignitions could cause the main composition to inflame and spoil the required effect.

FIG. 7.2 Cambric first fire (X50)

Slag Producers

The third group with the hottest outputs is that of slag producers. They are excellent igniters and are employed to give ignition in difficult situations such as at altitude where reduced pressure affects burning rates. They may be formulated either from fuel elements with salt oxidiser or from thermite maerials, fuel element with a metal oxide. Table 7.6 gives examples of this group of first fire compositions.

TABLE 7.6 SLAG PRODUCING FIRST FIRE COMPOSITIONS

Composition	*Uses*
$B/KClO_4$/thiokol	General priming
B/KNO_3	High altitude flare ignition
$Mg/BaO_2/C$	Tracer ignition
$Al/Si/KNO_3/Fe_3O_4/C$	HC smoke ignition
$Si/PbO_2/Cu_2O$	Phosphorus ignition
$Si/Ti/Pb_3O_4$	Delay ignition

Delays

Uses

The manufacture of delays is one of the prime uses of pyrotechnic compositions and has led to the development of theories for pyrotechnic combustion rather than the empiricism which has always surrounded pyrotechnics. In its simplest form, a delay can be used for operator safety, such as in safety fuze, which is used in blasting where accurate timing is unnecessary. However, some delays are required with great accuracy of burn time for use in delay detonators or to time function operations in sophisticated missile systems.

Timing Consistency

There are many factors affecting the design of a pyrotechnic delay. Perhaps the most important of these is the required time interval and acceptable limits. Delay compositions are known with burn times from one hundredth to forty seconds per linear inch. Thus desired timing can be achieved by choosing the appropriate composition to give a burn rate consistent with the space allowed to accommodate the delay column. For consistent timing, the column length should always be greater than its diameter. In fact, the volume and shape of the space available can affect delay design. Delays are often prepared in lead tubes which, being flexible, can be bent so as to fit the available space, for instance, into a spiral configuration. There is, however, a problem over heat removal since low gassing compositions have a temperature coefficient of 0.08 per cent °F (US data).

The Effect of Gassing

Also of paramount importance is the choice between compositions liberating large volumes of gas, such as gunpowder, and those which chemically are considered gasless. The gassy compositions must be vented to prevent pressure build-up, otherwise the burning rate would change as pressure changes. However, even the so-called gasless compositions are not truly gasless. Consider the reaction 7.4 where the reactants and products are all solids at normal temperatures. This reaction will produce a temperature

$$PbO_2 + Si \longrightarrow SiO_2 + Pb \qquad (7.4)$$

of over 2000°C and at this temperature there will be a high vapour pressure of lead; thus it is not a gasless reaction. The more gassy a reaction, the more dependent will be the burning rate on pressure. A completely gasless pyrotechnic probably does not exist although those approaching it are normally used in a closed or obturated system.

Horses for Courses

The choice of composition will depend mainly on the time interval required and whether the device is most suited to a vented or closed system. The most energetic compositions will give the fastest burn times and will be used in millisecond delays. Lower outputs will lead to longer time interval delays. Table 7.7 gives examples of delay compositions.

TABLE 7.7 DELAY COMPOSITIONS

Composition	Delay
PbO_2/Si $Mo/KClO_4$ $BaCrO_4/PbO_2/B$	Millisecond delays
BaO_2/Se $Fe/KMnO_4$ $Sb/KMnO_4$ $B/BaCrO_4$ $KNO_3/C/S$	Second delays
$W/BaCrO_4/KClO_4$/Superfloss (40 sec in^{-1})	Long delays

Great care must be taken over quality control of the ingredients and mixtures for delays, since changes in particle size and stoichiometry can make great differences in burning rates. Other factors affecting burn time are confinement, heat losses through the sheathing, and impurities.

Fire Starters and Incendiaries

If pyrotechnic compositions are to be used as fire starters, care must be taken in choosing a suitable oxidiser. If potassium chlorate or perchlorate is used, high temperatures will be achieved. However, one of the products, KCl, is an excellent fire-proofing agent and would likely extinguish the fire once started. Also, to effect ignition, a steady heat source is required not a very large flash of energy. This is best supplied by the so-called napalm which is a gelled fuel, such as kerosene; it is gelled by a mixture of aluminium naphthenate and palmitate, two soap-like molecules. Napalm is really the gelling agent but the word has been accepted in common parlance as referring to the thickened fuel. It is not really a pyrotechnic but due to its use as an incendiary is often considered as one. This type of material has been used for large-scale destruction as a bomb filling but is also used in small fire starters to aid survival in difficult circumstances. True pyrotechnic incendiaries have been employed in the small incendiary bombs used, to great effect by protagonists in World War II. Compositions such as $Al/Fe_3O_4/Ba\ (NO_3)_2$/boric acid are designed to burn and not explode and are required to ignite the bomb casing constructed from magnesium alloy. These are often called 'thermate' compositions.

Grenades filled with thermate materials are excellent for damaging metal structures such as large pieces of ordnance, making them inoperable and thus being useful for sabotage purposes. For example the American AN-M14 grenade penetrates one inch (2.5 centimetres) of mild steel in less than 30 seconds. Even higher performances have been noted for the device called the 'Pyronol torch' used by salvage crews for metal cutting in difficult circumstances. The basic composition was $Al/Ni/Fe_2O_3$ but more recently copper ferrite has been used increasing the output. It is thought that, on burning, gaseous iron is formed which acts as the cutting agent. Pyronol torches can cut through four inches of stainless steel at a depth of 2000 feet in a fraction of a second.

Smoke Production (Non-Screening)

The production of an aerosol of a dye or chemical agent by pyrotechnic means requires a composition which provides heat to vaporise the material; on cooling in the air, it condenses back to a cloud of small particles. The main problem is to provide enough heat for vaporisation but not to pyrolyse and destroy the dye or chemical agent since these are organic molecules. Also the heating composition itself should not produce smoke which would adversely affect the quality of the required aerosol or leave a residue which might impede release of vapour. Since low ignition temperatures must be used, the oxidiser is invariably $KClO_3$. The fuel may be lactose, usually as a 50/50 mixture, with oxidiser mixed with the dye or chemical agent. The riot control agent CS is an example. The other fuel used is sulphur which also requires the presence of sodium bicarbonate ($NaHCO_3$) to stabilise this potentially dangerous mixture. Two typical coloured smoke compositions are 25 per cent lactose/25 per cent $KClO_3$/50 per cent dye and 10 per cent S/25 per cent $KClO_3$/25 percent $NaHCO_3$/40 per cent dye.

In military use, coloured smokes are used for ground to air and ground to ground signalling. They are particularly useful for the marking of landing or dropping zones as they are good wind speed and direction indicators. They are also employed in search and rescue roles. Detail of a typical coloured signal smoke grenade is shown in Figure 7.3. The riot control agent, CS, has several delivery systems among which are cartridges for special guns, usually around 38 millimetre calibre, and hand thrown grenades similar to the coloured smoke grenade. Often barrel launched grenades will have sub-munitions which scatter on landing thus making them difficult to throw back.

Battery Activators

Modern weaponry is becoming increasingly dependent for electrical energy on batteries which have very long storage lives. One answer is to use a KCl-LiCl eutectic as the electrolyte which is solid at room temperature and

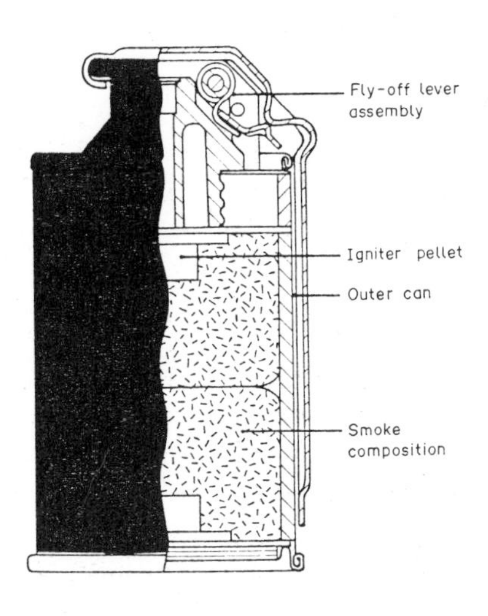

FIG. 7.3 Schermuly Signal Smoke grenade - details

thus the battery is inactive. In order to be activated, the electrolyte is melted by use of a pyrotechnic composition. This composition is usually a $Zr/BaCrO_4$ mixture impregnated on to silica paper forming a product known as heat paper developed by the Catalyst Research Corporation in the United States. By varying the stoichiometry and subsequent loading on the support, highly accurate heat outputs per given area of paper support can be achieved.

Self-heating Food Cans

Self-heating food cans were used extensively in the past, particularly by special forces, to provide hot food in difficult conditions. The pyrotechnic composition to be used must be relatively cool burning, gasless and it must not expand. The mixture generally used is 50 per cent $CaSi_2$/50 per cent Fe_3O_4 with a $CaSi_2/Pb_3O_4$/clay composition as first fire, held in a central channel within the can, Figure 7.4.

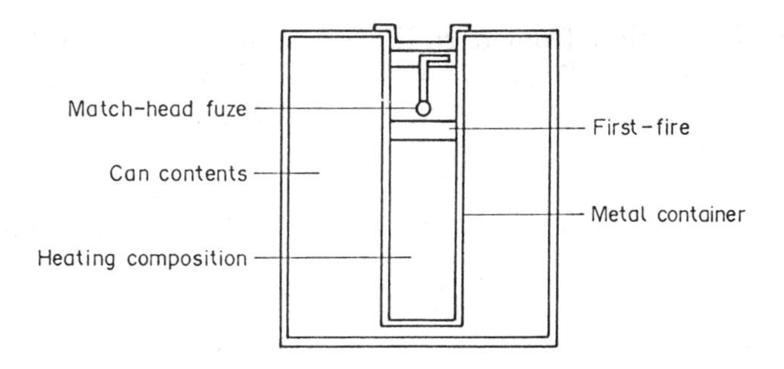

FIG. 7.4 Self-heating food can

Light (Electromagnetic Radiation)

Units of Measurement

Before considering pyrotechnically produced light, it is necessary to understand the rather complex units of measurement employed to quantify performance. To cover all types of electromagnetic radiation, and not just visible light, it is usual to assess energy outputs. Thus, the total energy emitted in all directions by a source is called the radiant energy; it is measured in watts seconds. The rate of energy production is the radiant flux; it is measured in watts. Radiant energy and radiant flux describe energy production in all directions. However, it is often necessary to consider the flow of energy in a particular direction. Radiant flux emitted from a surface element is called the radiant emittance (watts m^{-2}) or if from a point source, then radiant intensity (watts sr^{-1}) is used. The steradian (sr) is the unit solid angle: there are 4π steradians in a sphere. The radiant intensity can be found by measuring the total energy over a given area of an imaginary sphere surrounding the point source in watts. The solid angle producing that area is simply the area divided by the radius squared. Radiant intensity is then energy divided by the solid angle producing it (watts sr^{-1}).

When visible light is considered, more specific units are used. Luminous intensity, which is perceived brightness in a given direction, is measured in candela (cd) whereas the luminous intensity or luminence, depends on the area of the burning surface and is quoted as cd cm^{-2}. The rate at which the luminous energy is produced is called the luminous flux, measured in lumens (lm), where 1 lm is the luminous energy radiated per second per steradian by a uniform source of 1 cd luminous intensity. Specific luminous flux then relates output to the mass of composition and is given in lm g^{-1}. The basis for these units, the candela, is based on the output from a black body radiator of given area at a given temperature (2042K). To relate luminous units to direct energy output is complex since this output will be dependent upon the wavelength studied. For example, 1 watt is equal to approximately 600 lm at a wavelength of 555 millimetres (green light). Averaged over the whole visible spectrum, however, the value is around 200 lm W^{-1}.

The energy of emission from a radiating source is highly temperature dependent. For a perfect radiator of energy, a black body, energy emission is proportional to temperature to the fourth power (Stefan-Boltzman law). For broad spectrum emission, strongly exothermic compositions are required, forming solid products. For white light, the best combination is undoubtedly $Mg/NaNO_3$, which forms magnesium oxide (MgO) in the products. This combination can produce flame temperatures of over 2500K. Other outputs come from species present in the flame which would not exist at lower temperatures. These are excited ions or molecules which emit at specific wavelengths. One of them is important in the production of white light; it is NaH^+ which requires a source of hydrogen in the composition. This is why flares containing a binder have better outputs than those without one.

Illumination

Flares

Illumination of the ground is usually obtained by a flare candle burning in cigarette fashion from the base and suspended on a parachute. It may be deployed from aircraft or projected as mortars or shells. Figure 7.5 shows a sectioned aircraft five-inch recce flare. The compositions used are made from $Mg/NaNO_3$/binder with intensity and burn time being optimised for the particular end use by choosing a particular combination. For large area illumination millions of candela for hundreds of seconds are required, such output is achieved by the Bofors Super Lepus flare. The Schermuly helicopter flare for reconnaissance or emergency night landing provides 180,000 candela for 80 seconds from a payload of around 0.5 kg. A trip flare from the same company to identify infiltrators, provides 60,000 candela for 45 seconds from just over 1000 grammes of composition.

Photoflashes

To provide photoflashes for aerial photography at night much greater outputs are required but only for a very short time. The original photoflashes were true pyrotechnic compositions based on $Ba(NO_3)_2/Mg/KClO_4$. However, they were unsafe since they were invariably initiated to a high order explosive event by bullet impact. The explosion would have an equivalence of 0.5 kg TNT/kg photoflash. The outputs peak at several hundred million candela in about 10 milliseconds falling to one-tenth peak output by around 80 milliseconds. The camera is activated by a photocell with the shutter open during peak output. Modern photoflashes are not really pyrotechnics since they consist of a finely divided metal powder or flake. This is usually aluminium, which is burst into a cloud and simultaneously ignited by a central high explosive burster charge. Five hundred million candela peak intensity can be produced.

Signalling and Tracking

Signalling and tracking flares are usually coloured by adding materials to the flame which emit light energy at specific wavelengths to produce colour in addition to a white continuum produced by the other products. Green light (500-550 nm) is provided by the presence of $BaCl^+$ in the flame whereas red light requires emission in the range 650-750 nm given out by $SrCl^+$. The compositions normally employed contain magnesium as fuel along with two oxidisers: potassium perchlorate provides the chlorine and either strontium or barium nitrate accompanies it.

The outputs of coloured lights are usually much lower than those found for illumination compositions. This is due to the fact that the white light background can wash out the colour if it is too intense. This is not a major

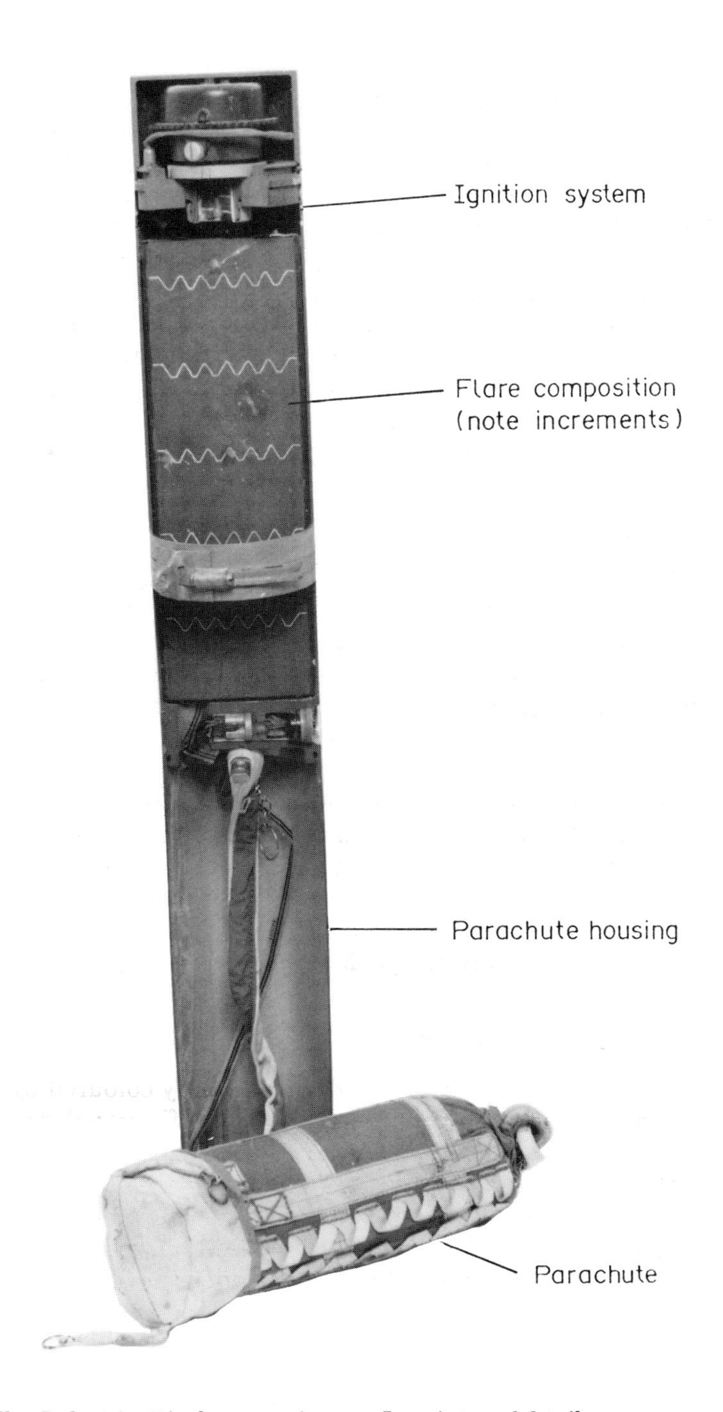

FIG. 7.5 Wallop Industries 5 inch reconnaissance flare, internal details

problem for red or green signals since the human eye is quite sensitive to these colours. However, blue flares are not used since the eye is relatively insensitive to the colour blue. The luminous intensity is normally less than 50,000 candela. Coloured flares for signalling may be fired either as stars or with a parachute. There are also hand-held devices particularly for search and rescue operations. They are designed to be readily distinguished at some distance.

Tracking flares, however, have some more unique problems. They may well operate at high altitude, and therefore reduced pressure and be set in the tail of a supersonic missile. The output must also be matched to the likely luminosity of the background which may be a bright sky. Several specialist compositions have arisen from these requirements, for example Ti/Sr $(NO_3)_2$, which also reduces smoke from the pyrotechnic. This is used in the Rapier tracking flare, Figure 7.6.

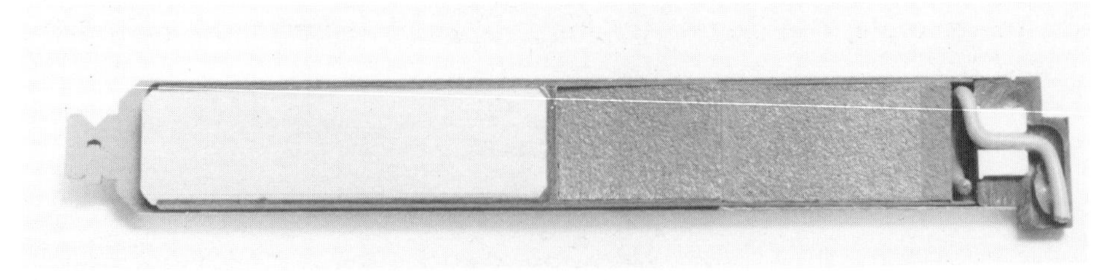

FIG. 7.6 Tracking flare, internal detail

A further type of tracking composition is that used in tracers for shells and small arms. The spin imparted to these projectiles and their high velocities further complicates the job of the pyrotechnician. A typical spin rate for a small arms round is 400,000 rpm with an acceleration of around 200,000 *g*. Early tracer compositions had problems of propagation failure or burning out too quickly. Modern tracers use compositions which have solid slag-like products to hold the heat and aid propagation. It is also usual to have a dark ignition mixture pressed on to the light producing composition, to prevent blinding those shooting and not give away their exact location. The tracer mixture must have a short burn time with high intensity. It is common practice to use magnesium as fuel with a metal peroxide as oxidiser, BaO_2 (green), SrO_2 (red). A PVC binder is often added.

Decoy Flares

Many modern missiles, anti-aircraft and anti-ship in particular, operate on an infra-red homing guidance system. To counteract this threat, infra-red flares have been developed to be fired to distract the incoming missile. Such decoy flares require rapid rise times to maximum output and very high output values. Take, for example, an aircraft decoy flare. It must appear as a more intense heat source than the exhaust of a jet engine. To achieve this very high energy, compositions are required which are highly energetic and have much of their output in the infra-red region of the spectrum. At this time only one composition fills the requirements, a mixture of powdered magnesium, teflon and viton. The oxidiser in this composition is the organic polymer polytetrafluoroethylene (PTFE) which reacts with the magnesium to produce MgF_2 and around 10 kJ g^{-1} of energy.

The composition is often pressed to produce the candle but it can be extruded. The standard devices for these flares are rectangular and are denoted by their size in imperial units (inches). Two sizes are commonly used for aircraft, 118 and 218. An example of the latter which measures 2 x 1 x 8 inches is shown in Figure 7.7. The cutaway shows the impulse cartridge to eject the flare, and the firing unit which ignites the flare candle as it leaves the container. This flare reaches a maximum output of 20 kW sr^{-1} in around 0.5 second and burns for about 4 seconds. Other shapes are used: for example, the Royal Air Force Tornado carries the cylindrical WI Cart CM55.

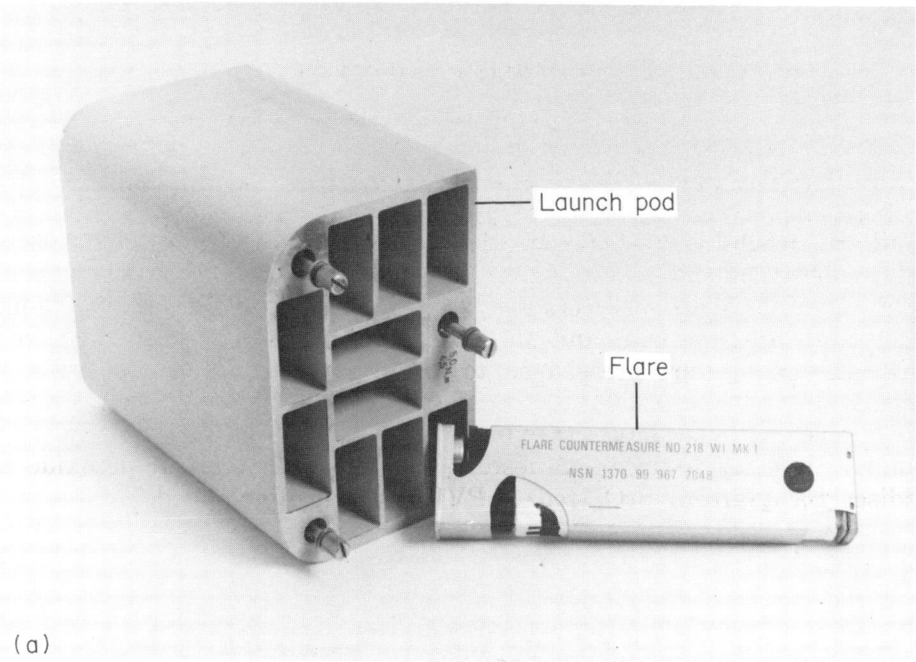

(a)

FIG. 7.7a Launch pod

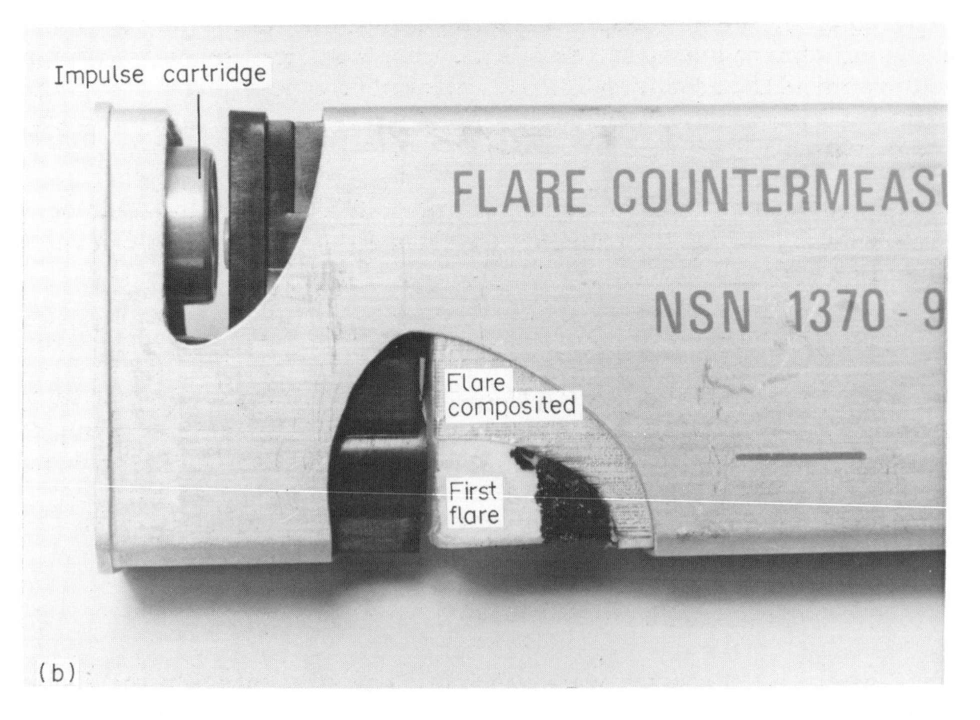

FIG. 7.7b '218' infra-red flare

When used to protect ships these flares are used in conjunction with chaff to create maximum confusion. In this role they are often suspended by parachute to give a relatively fixed emission point.

Screening Smoke

The Nature of Smoke

A smoke is a dispersion of particles in the atmosphere which settle under gravity very slowly due to their small size, typically in the range 0.1 to 5 microns. As already discussed in the section on the uses of pyrotechnic heat, the production of coloured smokes for signalling is achieved by the volatilisation of a dye which recondenses to small solid particles in the atmosphere: this produces material suitable for signalling. To give a screening effect, other methods must be employed. These range from burning pyrotechnic compositions to mechanical pulverisation and ejection of the powder thus formed. The physical methods are outside the scope of this book and will not be considered further.

To understand the performance of screening smokes, it is necessary to understand the physics of obscuration. Any observation system, whether electronic or the human eye, is assessing contrast, that is the difference between irradiance falling on the detector from the target (R_t) and that from the background (R_B); equation 7.5. If a smoke is introduced

$$\text{Contrast} = t\ R_t - R_B \qquad (7.5)$$

between target and detector, then the contrast is reduced by a factor τ, equation 7.6, where τ is the transmission through the cloud and has a value between 1 and 0.

$$\text{Contrast} = \tau\ (R_t - R_B) \qquad (7.6)$$

Obscuration is achieved when the contrast falls below the threshold value for the detector. Smokes are usually assessed by their attenuation or extinction factor $(1 - \tau)$ and this is normally expressed by the mass extinction coefficient (α) which relates to τ, path length through the cloud (L m) and concentration of smoke particles in the atmosphere (C g m^{-3}). The relationship is given in equation 7.7 and for absolute values

$$\alpha(\lambda) = [-(\ln\tau)_{\lambda}]/CL \text{ m}^2/\text{g aerosol} \qquad (7.7)$$

should be quoted at a given wavelength (λ). The attenuation in the visible region is created by light scattering by the particles and in some cases a simultaneous absorption also occurs. Of importance to these processes is the particle size and shape.

Production of Smoke

The production of visual screening smokes by pyrotechnic compositions relies on the formation of products which can absorb moisture from the atmosphere. It often undergoes further chemical reactions in the process. A good example of this, although not strictly a pyrotechnic composition, is phosphorus, which is often used to produce screening smoke. In this case the phosphorus burns in the air to form oxides of phosphorus. These in turn react with atmospheric moisture to form phosphorus based acids which absorb more moisture to give liquid droplets creating the smoke. An added bonus in this case is that there is an increase in mass such that the yield factor (Y), the ratio of smoke weight to composition weight is greater than 1. Mass extinction coefficient values for phosphorus smokes of 6 m^2 g^{-1} are possible. More conventional compositions are based on hexachloroethane (C_2Cl_6) and zinc oxide (ZnO) with fuel such as aluminium or calcium silicide ($CaSi_2$). They produce the smoke by forming zinc chloride ($ZnCl_2$) which again picks up moisture to give a hydrated system: (it also increases particle size, giving mass extinction coefficient values of around 4 m^2 g^{-1}.) For these types of screening smoke, the relative humidity in the atmosphere is critical to the quality of the smoke. For example, with a phosphorus based smoke around twice as much composition may be required to give the same effect at 40 per cent relative humidity (RH) as a given weight at 100 per cent RH.

Phosphorus smokes can be produced either by a cartridge with a central burster, usually gunpowder, or by mixing the phosphorus with a heat-producing composition. It vaporises the phosphorus which subsequently burns as it reaches the atmosphere from a vent hole. The hexachloroethane based systems are pressed into containers and burn either cigarette fashion or down a central channel. Aluminium is normally used as fuel, although cool burning compositions using manganese have been formulated to eliminate flame.

Infra-Red Smoke

Screening in the infra-red region (3 - 14 μm) is far more difficult and compositions which are good visible screeners are much less effective in this region. For example, the mass extinction coefficient (α) for a hexachloroethane smoke may reach 5 m^2 g^{-1} in the visible region, however, the maximum value in the infra-red region is around 0.1 m^2 g^{-1}. One answer may be to increase the smoke concentration, although this may provide a severe logistics burden since concentration would need to be cubed to screen in the far infra-red. It would be preferable to find materials with higher α values in the infra-red region. This has led to non-pyrotechnically generated aerosol clouds of materials, such as metal flake with α values around 1 m^2 g^{-1}.

Short Term Screening

To obtain a satisfactory screen, using burning materials, is possible if the emission also contributes to the screening effect. This can be achieved using phosphorus bursting munitions in a pattern around, say, an armoured fighting vehicle, such as a main battle tank. To achieve rapid screening, some or all of the munitions must airburst a short distance from the vehicle and in line with the threat. Screening of this type is relatively short-lived and it is likely that long duration infra-red screening will not use pyrotechnic means. Research continues around the world to develop screening smokes which will be effective over wavebands from 0.5 μm to around 1 centimetre.

Sound

Producing Noise

Noise from any explosion is caused by a pressure wave in the air. These waves travel at sub-sonic speeds. They are called acoustic waves and, in the case of a detonation, are degraded shock-waves. However, for pyrotechnics, the acoustic wave is caused by a sudden release of high pressure gas. This can be achieved in two ways. First, relatively slow burning compositions, such as gunpowder, are ignited within a stout container, which is usually manufactured from paper and glue to eliminate fragmentation. This method relies on the burning rate accelerating due to the pressure rise within the container. This causes the container to rupture and release gas. The second method is to choose a very fast burning composition which can produce the gaseous products so quickly as to produce a pressure wave with light confinement or in some cases unconfined such as Pb_3O_4/Ti.

Uses of Noise

The choice of composition depends to a large extent on the use to which the noise is being put. Noise-making stores have two major uses, signalling and training aids for battlefield simulation. Some training aids do find a third use which, although minor, is nonetheless worthy of mention. It is in the area of internal security where hostage release requires use of minimum force. Here, distraction and disorientation can be achieved by the use of devices called stun grenades by the popular press. These grenades are a variation of the training grenade containing a noise making composition.

Compositions

Particularly in the area of battlefield simulators, the compositions must be carefully tailored to match the apparent sound from the actual weapon. Some simulators contain only gunpowder; however, it is more usual to have more energetic systems. The usual compositions for flash and sound are based on either magnesium or aluminium or both together with potassium perchlorate and sometimes antimony sulphide Sb_2S_3 is added. Another series of compositions from the United Kingdom is based on ammonium nitrate and paint grade aluminium. Some of these contain TNT and they are all initiated by a detonator thus putting these compositions on the borderline between pyrotechnics and high explosives. This latter series gives good simulations for the larger weapons, such as mortars and large guns.

Dangers

The high energy rapidly burning compositions are amongst the most dangerous of pyrotechnics. Accidents have caused serious injuries when pyrotechnics have exploded in air; and when they have exploded underwater near divers, they have caused death. The pressure wave from this type of store is extremely dangerous underwater. Even so it is common practice to signal to clearance divers by dropping a thunderflash into the water.

Whistling Compositions

Another fascinating pyrotechnically generated noise is the whistle. It has long been known that certain compositions emit a loud whistle when burned in tubes. They are based on aromatic (benzene based) acids or their salts, together with a suitable oxidiser. Fuels which have been used to include, gallic acid, potassium benzoate, picric acid and sodium salicylate. The oxidiser may be potassium chlorate or perchlorate and also potassium nitrate.

The whistle effect is caused by a regular intermittent burning at the surface of the composition due to the crystalline nature of the fuel. The shattering of the crystals as the burning surface proceeds causes the regular speeding up, then slowing, of the burning rate at a set frequency, thus producing a tone. The whistle so produced can be employed in simulators, particularly for shell bursts, in which the whistling sound simulates the shell flight before the simulated burst. Again, a note of caution must be given with these compositions, since they are capable of high-order detonation in some circumstances.

8. Pyromechanisms

Categories

The title, pyromechanisms, covers a wide range of devices which fall broadly into the following categories:

- ▷ Mechanical actuators
- ▷ Gas generators
- ▷ Explosive separators (using detonating explosive fillings).

They find their major uses in aerospace, aircraft, missiles, torpedoes and the off-shore oil industry. The advantages associated with explosive-actuated devices are that they have high power-to-weight ratios and can thus be relatively small, together with high reliability and very short operating times.

Explosive Fillings

Pyrotechnics (See also Chapter 7)

Mechanical actuators operate by the production of hot gas and can be powered by almost any type of explosive material including those which detonate. However, it is usually the practice to use an explosive which only deflagrates. For pyromechanisms they fall into three categories. The first is the true pyrotechnics. Fuel/oxidiser mixtures such as black powder, often without the sulphur to prevent corrosion problems, are utilised. Other pyrotechnic compositions used are aluminium or zirconium with potassium perchlorate.

Single Compounds

The second type of explosives used is single compounds. A family of compounds of this type are the lead salts of mono-, di- and trinitroresorcinate, known as LMNR, LDNR and lead styphnate, respectively. A recent development has been the introduction of potassium dinitrobenzofuroxan (KDNBF). These materials are all sensitive primary explosives and are found both as the first layer in a multi-component system or in very small devices as the only explosive present.

Propellants *(See also Chapter 5)*

The third category of explosives found in mechanical actuators are propellant compositions. The types used are, single base containing nitrocellulose and double-base having both nitrocellulose and nitroglycerine content. The choice of material and its grain size and shape depends upon the type of output required. For rapid pressure build-up, powders are used with the individual grains being either spheres or flakes, the most rapid deflagration being obtained with flake. As grain size increases the burn time is extended giving a more prolonged thrust, probably with lower peak pressure. For uses with a requirement of high accuracy double base propellant is preferred as it is the more reliable in its initiation and in quality of manufacture.

Gas generators will be normally powered by a propellant: they are usually of the double-base type in the form of large extruded grains. This provides large quantities of very hot gas for many seconds, for example, to provide thrust to start a turbine. If, however, the gas is required for inflation devices the gas must be relatively cool. This can be achieved by a physical process to cool the hot gas from a double-base propellant or by choosing a very cool burning composition. An example of this cool burning type is guanidine nitrate and manganese dioxide (90/10) claimed in a Japanese report to burn at 200°C.

Detonating Explosives

Detonating explosives are utilised by pyromechanisms where a more shattering force is required than can be obtained by burning explosives. The primary explosive of choice is usually lead azide which is readily initiated by many stimuli. The main charge is often an explosive which would be considered as a booster in more general explosives application. This is because the pyromechanism will be quite small and thus the problem of critical diameter must be considered. Explosives are required which will still detonate when present in containers down to a few millimetres in diameter. The two materials most often used are pure PETN and RDX/wax; the latter has only a very low percentage of wax to make the composition a little less sensitive than pure RDX. For larger devices tetryl may be used. However, as the device becomes larger it is debatable whether they are still pyromechanisms.

Initiation

Initiation in general terms has been described in Chapter 4. For pyromechanisms the initiation stimulus is usually electrical and devices thus initiated are known as electro-explosive devices (EEDs). A low energy source is often used to heat a bridgewire which may be embedded in a first-fire layer of composition or be present as part of a fuze-head. To ensure safety, the all-fire and no-fire conditions of such devices should be known so as to choose the correct bridgewire to match operating conditions. For example, a typical EED bridgewire igniter would have a resistance of 1 Ω, a minimum all-fire current of 3.5 amps, a firing current for actual use of 5 amps and a no-fire current of 1 amp for five minutes. The no-fire conditions must take into account the possibility of induced current within the firing circuit. This problem can be alleviated to some extent by using a radio-frequency filtering device to prevent induced currents and allowing a lower firing current to be used. Alternatively, if the device uses a high explosive fillng, an exploding bridgewire initiation can be used with no possibility of accidental initiation by induced current.

Many other methods may be employed for initiation of pyromechanisms including stab and percussion. Due to the specialised nature of these devices some more unusual methods have been employed. For example lasers have been used to initiate power cartridges, the beam being carried to the device via fibre optic links. Another less common method is to use a detonation wave from a donor charge passing through a bulkhead to initiate a deflagrating system on the other side. Thus transfer systems can be high explosive, so they are very rapid and the through-bulkhead initiator steps down from a detonation to an igniferous system. There is also an added advantage that any likelihood of a blow-back through the device is eliminated. These units withstand pressures as high as 50,000 psi.

Devices

Mechanical

Mechanical actuators are usually piston devices in which the firing of the charge moves a piston within a barrel. Figure 8.1 gives some examples of pyromechanical actuators. For an actuator the explosive composition is usually in the body of the device. For larger devices it is usual to have the mechanical part as a unit to which is screwed a power cartridge. These are often called thrusters or pullers if the action is a push or a pull. Alternatively, the power cartridge can drive a guillotine blade to cut a cable, as in Figure 8.1, or wire or move a blade to close or open a valve. An undersea cable cut by a pyromechanical cable cutter using 60 g of propellant is shown in Figure 8.2 and a guillotine bolt cutter used to release torpedo nose covers for recovery in testing in Figure 8.3.

From Figure 8.1 it can be seen that a piston as such is not always necessary. In the dimple actuator and the bellows actuator, the actions require no piston. One interesting advantage of the bellows actuator is that it is able to turn corners with the aid of a guide.

Actuators can be used to operate switches, activate release mechanisms such as shaft locks in torpedoes, or harness release for ejector seat systems and numerous other applications. The small devices may simply contain an electric bridgewire or fuze-head with either black powder or a deflagrating compound such as LDNR. As the size of the device increases, the move is towards a propellant filling which can be either single or double-base.

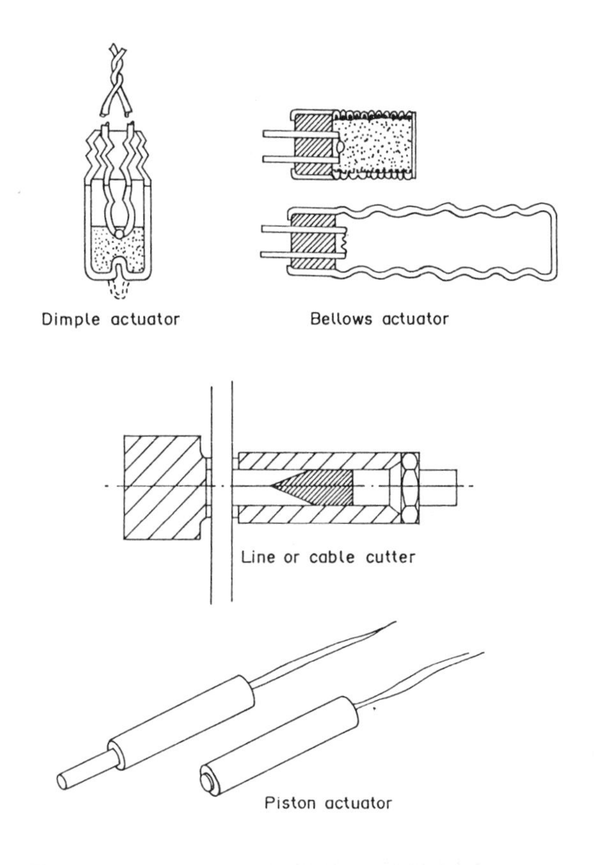

FIG. 8.1 Pyromechanical actuators and thrusters

Gas Operated

Gas generators are usually powered by a large single grain of double-base propellant, burning either on the inside of a central conduit or cigarette fashion. The gas efflux can be used to spin up a gyroscope for missile guidance systems. Another use for gas generators is to provide pressure on piston surfaces to drive liquid fuel from storage tanks into the combustion chamber of a rocket motor, The United States Lance missile, in Figure 8.4, uses this system.

Detonating

Examples of the use of detonating systems are surprisingly quite numerous. Since the explosive loading is usually small the damage from the

FIG. 8.2 Undersea cable cut for a guillotine cutter

detonation can be contained or controlled. For instance, the transfer system called Confined Detonating Fuze (CDF) has an explosive loading of at most 1 g m^{-1} contained in a metal tube, made of lead, aluminium or silver, which is covered with nylon and fibreglass braid. All the products from the explosion are retained within the structure. If containment is not required, the device is called Mild Detonating Fuze (MDF). This may be used to create specific damage to a structure. The zig-zag pattern seen on some aircraft canopies is MDF whose role is to shatter the canopy prior to seat ejection.

Rapid release can be achieved in many circumstances by employing explosive bolts, nuts, and shackles. Again a detonating explosive is used. Its brisance shatters the structure, usually at a pre-weakened site.

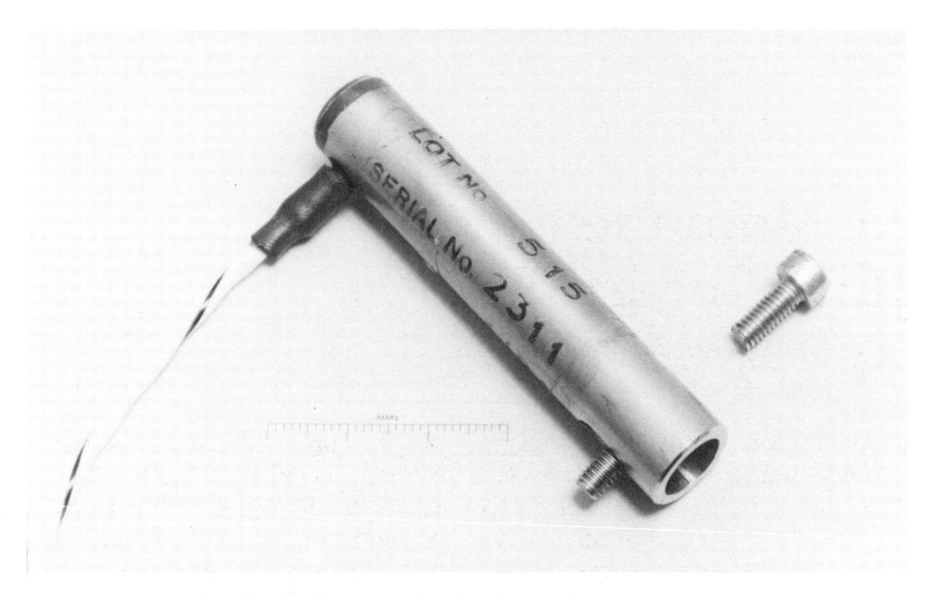

FIG. 8.3 Guillotine bolt cutter for a release mechanism

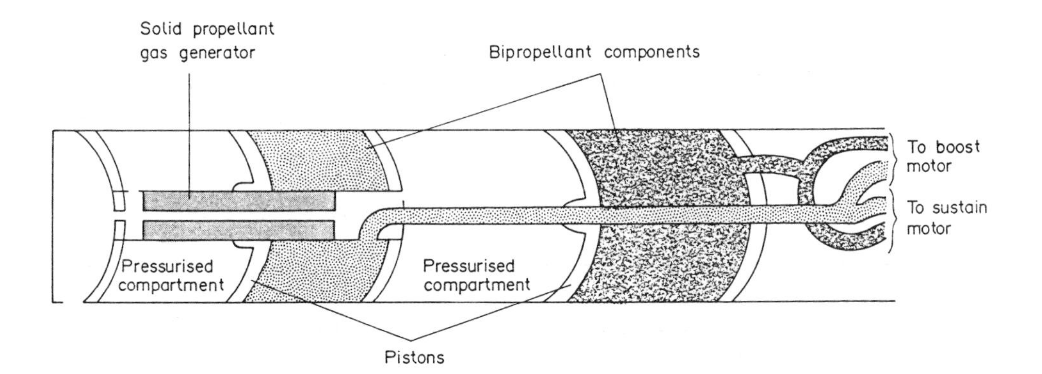

FIG. 8.4 Gas generator in the Lance missile

9. The Manufacture and Filling Integrity of Explosive Charges

Explosives and propellant charges may be produced by a variety of manufacturing techniques; those commonly employed in industry at present include melt casting, pressing and polymer bonding. The particular technique chosen is determined by the physical properties of the ingredients of the explosive formulation. The vulnerability of the finished charge to accidental initiation is in turn often determined by the production technique chosen.

Melt-Casting of HE Charges

Typical ingredients of HE formulations include explosives such as TNT, RDX and HMX, oxidisers such as ammonium nitrate and ammonium perchlorate and aluminium as fuel. The physical properties of some explosives are shown in Table 9.1.

TABLE 9.1 PHYSICAL PROPERTIES OF HIGH EXPLOSIVES

Explosive	*Melting Point (°C)*	*Ignition Temp. (°C)*	*Density* ($g\ cm^{-3}$)	*VOD* ($km\ s^{-1}$)	*FOI*
α-TNT	80.9	240	1.65	6.9	160
RDX	204.1	213	1.80	8.78	80
β-HMX	246	300	1.90	9.1	60
DATB	286	305			

TNT

The melting point of TNT is conveniently low so that it is readily melted by steam heat, yet it is high enough to be unaffected by the normally encountered environmental temperature range (-40 to + 70°C). Its ignition temperature is well separated from its melting point so that the large-scale melting of the explosive is a safe operation. Other high explosives melt at above 200°C, at which temperature decomposition of the explosives is beginning to occur. Furthermore, the ignition temperatures of the explosives are dangerously close to their melting points, so that the melting operation is not an inherently safe procedure. For these reasons TNT forms the basis for most melt-cast filling of munitions with high explosive formulations.

The melting point of commercial TNT ranges from 79.8°C to 80.6°C. The lowering of the melting point is due to the presence of impurities produced in the chemical preparation process. The impurities are useful; they plasticise the TNT and help to prevent cracking of the charge at low temperatures but they may give rise to exudation at high temperatures.

RDX and HMX

RDX and HMX cannot be melt-cast but when formulated into a slurry with melted TNT they may be successfully processed. Alternatively they may be cast in admixture with suitable waxes or cast, extruded or pressed into munitions when blended with plastic binders.

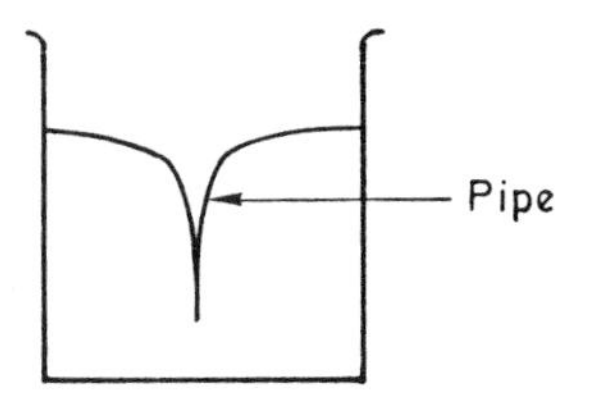

FIG. 9.1 The solidification of TNT

The Meissner Process

Molten TNT shrinks by some 10.8 per cent on setting. A vessel filled with molten TNT, loses heat through its walls so that the TNT solidifies at the vessel walls. When solidification is complete a hollow pipe is left in the centre of the charge (see Figure 9.1). Furthermore, any impurities in the TNT are concentrated in the last portion of the liquid to solidify; this material has a lower melting point and inferior strength to the rest of the charge. Many production methods have been developed to overcome this problem for TNT and TNT slurries of which the Meissner process is typical.

In this process, a header funnel is fitted to the munition to be filled so that extra melt to fill up the potential pipe may be accommodated. A hot finger heated by steam is inserted down the axis of the munition to the bottom of the melt (see Figure 9.2). The hot finger is attached to a screw withdrawal mechanism so that it may be removed slowly from the melt by a programmed withdrawal. As the finger is removed, cooling and solidification of the TNT-based melt occurs from the bottom of the munition. The controlled cooling allows the explosive to be annealed as it sets thus alleviating thermal stress and obviating the growth of planar TNT crystals which would serve as weaknesses for cracking. Eventually, the header funnel together with the now solidified excess explosive in which any impurity is concentrated is removed and is referred to as the 'lost header'.

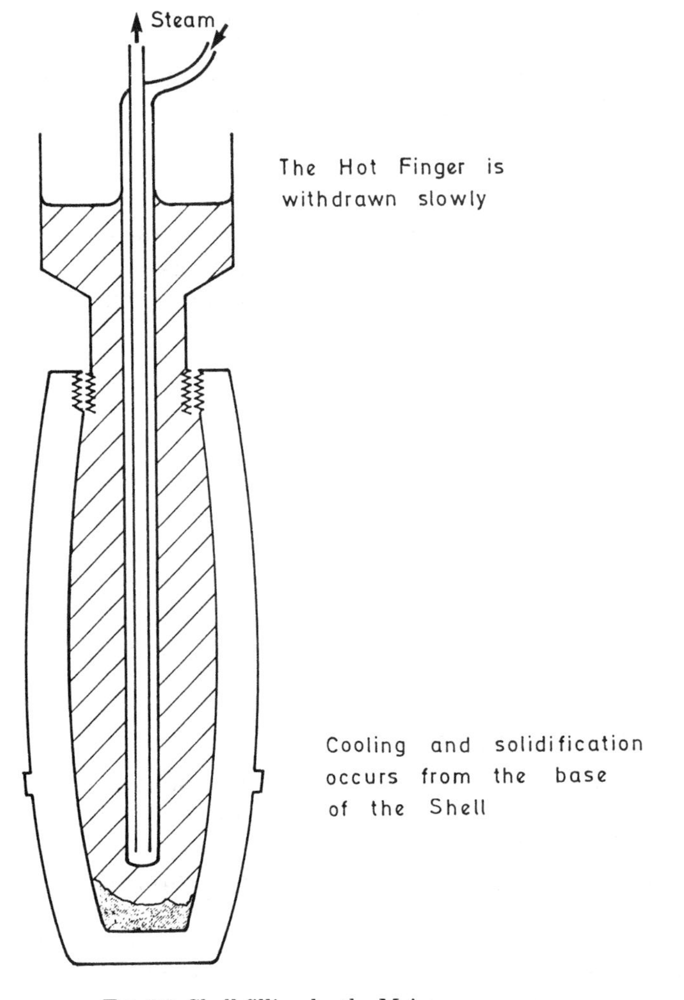

FIG. 9.2 Shell filling by the Meissner process

Typical TNT-based Melt Cast Formulations

Typical compositions include TNT, 60:40 RDX/TNT, Composition B, various Torpex formulations containing RDX, TNT and Aluminium, Octols containing TNT and HMX, and Explosives Department Composition No. 1. (EDC 1) containing HMX, RDX, TNT and beeswax.

It should be noted that Composition B and 60:40 RDX/TNT are not identical; the 'RDX' used in Composition B contains between 10 per cent and 16 per cent of HMX. RDX and HMX are slightly soluble in TNT (4 per cent approximately) and the sharper corners of the nitramine particles are dissolved away, permitting better packing of the solid components of the slurry.

Viscosity considerations limit the solids content of TNT slurries which may be poured and cast to about 74 per cent (70 per cent solids, 4 per cent in solution). Somewhat higher solids loadings may be achieved with vacuum casting processes.

The packing of solids is markedly affected by the particle size and shape of the solid. Spherical particles pack much more efficiently than other shapes. Powders with particles all the same size give poor packing however and practical explosives formulations require a polydisperse distribution of sizes so that coarse, medium and fine particles are present in the ratio 100:10:1. In this case, the smaller particles are able to occupy the interstices between the larger particles. Higher packing densities and hence better explosive performance can be achieved in this way.

The strength and resistance to cracking of TNT-based charges is determined to a large extent by the size and habit of the TNT crystals produced when the formulation solidifies. As has been indicated above, control of this is effected by the strict adherence to an optimum cooling programme. The TNT crystal size and habit may also be modified by co-crystallising small quantities of hexa-nitro stilbene (HNS) with the TNT. The ultimate strength of the TNT charge is increased by at least a factor of five by this method.

Some manufacturers use very crude methods to control the cooling and solidification process. One method is to insulate the upper part of a munition with a felt muff, leaving the lower part of the munition uninsulated. Whilst such simple methods may be satisfactory for some purposes, such munitions will show a greater frequency of defects on inspection.

TABLE 9.2 Advantages and Disadvantages of Melt Casting

Advantages	*Disadvantages*
Simple process	Charges prone to cracking
Cheap plant	Charges have high explosiveness and sensitiveness
Low unit costs	Settling of ingredients during solidification can lead to inhomogeneity
High volume production easy	

Pressing of High Explosive Fillings

Direct Pressing

The direct die pressing of explosives into a case results in very reproducible fillings, does not usually call for very high temperatures and is readily adaptable for automatic processing. An example of its use is in the production of shaped-charge bomblets for the MLRS system. Figure 9.3 illustrates such a scheme for the direct production of shaped charges: the pressures used may be as high as 70 MPa.

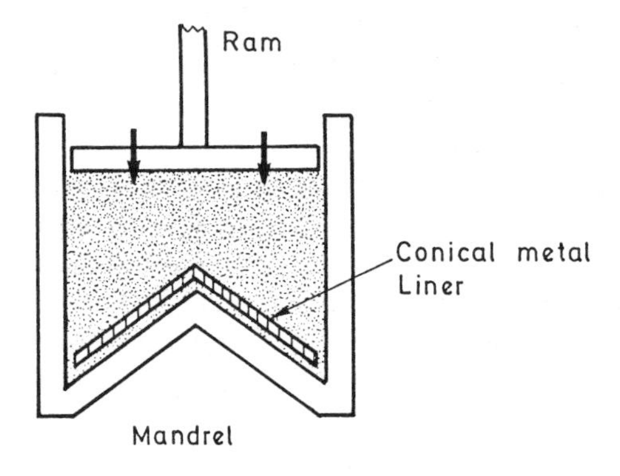

FIG. 9.3 Die pressing of explosives

Uniformity of charge density is excellent when controlled to a fixed pressure loading on the ram rather than to a fixed stop. To ensure reproducibility, the crystal grist size of the explosive mixture must be reproducible. It is possible to reach densities of 95 per cent of the crystal density, and, if vacuum pressing techniques are employed so that the air between the grains is not trapped, densities of 99 per cent of the crystal density may be achieved. Vacuum pressing is essential if the explosive undergoes plastic flow when pressed. This would result in the formation of gas bubbles throughout the explosive charge.

In spite of the convenience and reproducibility of the method, it is very expensive to develop and costs rise sharply as the permitted tolerances in the tools decrease. Other disadvantages include safety problems where explosive may be trapped in the clearances between the die and the ram.

When an explosive is pressed in a die or case by a ram, then friction at the walls causes pressure and density gradients within the charge. Additionally, one dimensional pressing is bound to give rise to anisotropy and residual strains in the fabricated charge. This leads to charge distortion as strains are gradually allowed to relax. The development of incremental pressing is an attempt to reduce the magnitude of these problems.

Incremental Pressing

Instead of fabricating a charge in one stroke of the ram, the explosive is added to the die and pressed after each addition. Figure 9.4 illustrates the principle. The effects of wall friction are greatly reduced but variations in the charge density are still extensive.

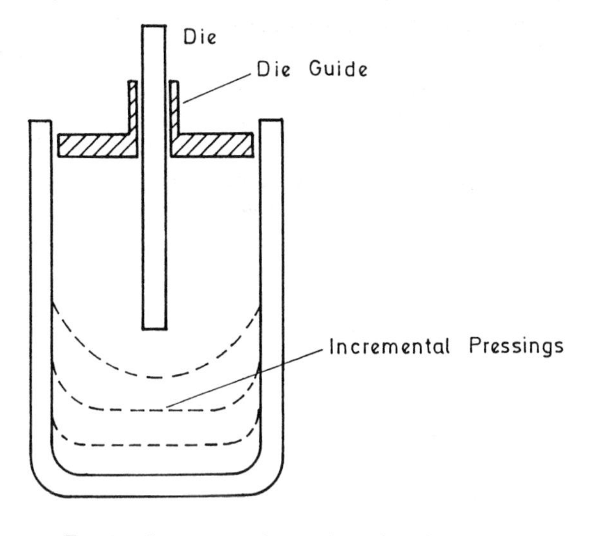

FIG. 9.4 Incremental pressing of explosives

Hydrostatic and Isostatic Pressing

Where dimensional stability, uniformity and high charge density are of paramount importance then hydrostatic pressing or, if this does not result in sufficient uniformity, isostatic pressing is used. The two processes are illustrated in Figures 9.5 and 9.6.

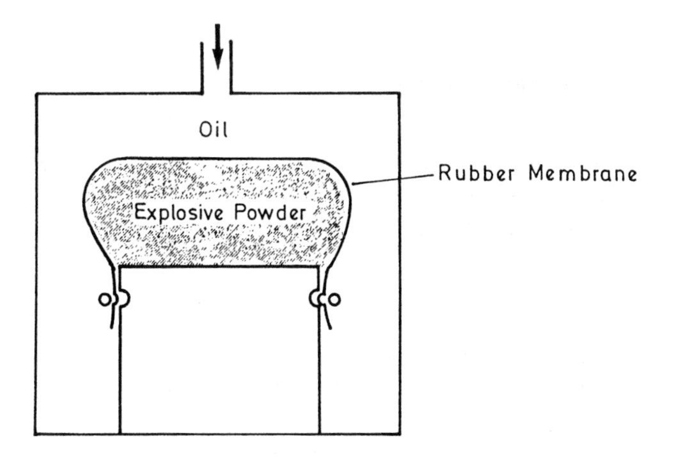

FIG. 9.5 Hydrostatic pressing of explosives

With these techniques the explosive is contained by a rubber membrane and evacuated before pressure is applied. Wall friction is completely eliminated by these techniques but, with the directionality of compression experienced by the explosive with the hydrostatic technique, there is still some anisotropy and residual strain in the pressed charge. With the isostatic technique, however, these problems are almost eliminated. The quality of the charge can be enhanced by pressing at high temperatures, typically about 120°C for some eight to ten hours. Typical pressures may be 135 - 200 MPa. The elimination of wall friction means that friction sensitive explosives such as RDX and HMX are amenable to this process.

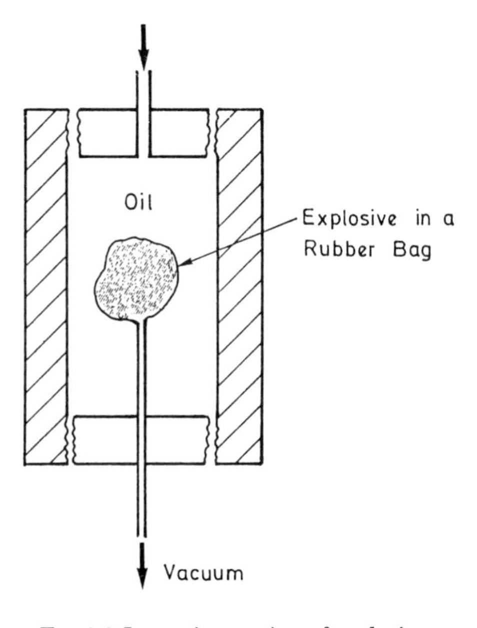

FIG. 9.6 Isostatic pressing of explosives

Pressure vessels to withstand such high pressures are expensive. Use is often made of gun barrels and since the largest calibre readily available is sixteen inches the maximum dimensions of the explosive charge are limited.

Machining

After pressing, the charge will require machining. All standard machine-shop operations such as turning, milling, drilling, boring and sawing can be carried out on the explosive. Any shape of explosive surface which may be required for various applications such as detonation wave shaping, metal forming or other specialised output can be fabricated with ease to an accuracy of 0.025 millimetre. The accuracy is limited by the dimensional stability of the pressed explosive. For safety, machining operations are carried out remotely and cutting edges are generally liquid cooled. It is an advantage to use as low a cutting speed as possible. Probably the most difficult task is to ensure adequate cooling during the drilling of small diameter holes. In all machining operations, care must be taken to collect all dust and swarf to prevent the build-up of an explosive hazard.

Such operations are expensive and can only be justified for high cost systems where performance is paramount and tolerances are tight. Applications are often to be found in the design of trigger systems for nuclear weapons.

Plastic-bonded Explosives (PBX)

The chief disadvantages of charges manufactured from TNT-based formulations are their propensity to crack when roughly handled, their inherent high explosiveness and sensitiveness, their potential inhomogeneity and their low temperature resistance. The upgrading of the requirements for modern military explosive charges has led to the development of plastic-bonded explosives (PBX) which have greater heat resistance than melt-cast explosives. They also have much improved strength, dimensional stability and cracking resistance. The temperature of the charge in a high velocity missile may rise to as high as 200°C during flight; meltcast explosives could not be expected to function reliably in this situation. Many ordnance systems are required to remain safe if dropped from a considerable height or if they are exposed to an accidental fire and again, TNT-based systems are not satisfactory.

The sensitiveness and explosiveness of charges has been found to be related amongst other factors to the shear properties and the strength of binding of the individual explosives particles into the bulk charge. The properties of PBXs may be more easily controlled and modified to successfully resist the kinds of environmental stress a charge may encounter than the older cast TNT charges.

Plastic-bonded Compositions

Most PBXs consist of a particulate explosive embedded in a continuous or binder phase. The particulate explosives may be produced by a variety of manufacturing techniques including precipitation, crystallisation, colloid milling and fluid energy milling. The formulation is then made up to form a slurry or paste if it is to be fabricated, using pour casting or extrusion casting processes. It is presented in the form of a moulding powder if the pressing processes discussed earlier are to be used.

The binders fall into two main classes.

- ▷ For casting or extrusion processing, binders which are cured *in situ* are used. Typical binders include polyurethanes, silicones and epoxy resins.

- ▷ For pressing operations, pre-cured binders such as Viton rubber, PTFE and KEL F are used.

In a cast charge there may be as much as 20 per cent of binder and for an extruded composition the binder content may be in the region of 12 per cent. A considerable improvement in explosive performance can be attained if the binder is itself an energetic material; current research is aimed in this direction. An improvement n explosive performance in charges fabricated with conventional binders may be gained if an energetic plasticiser is incorporated with the binder. The oldest and still used example of this is nitroglycerine. It has disadvantages however. It is difficult to use because it is very sensitive, thermally unstable and is toxic. Other more attractive energetic plasticisers are now available.

PBX Binders

The requirements of a suitable binder for a PBX are:

- ▷ Thermal stability
- ▷ Low toxicity
- ▷ Compatibility with the incorporated explosives
- ▷ Availability
- ▷ Ease of processing i.e. small number of reagents, low viscosity, room temperature mixing and long pot-life.
- ▷ Appropriate cure characteristics i.e. not violently exothermic and reasonably short curing period.
- ▷ Low glass transition temperature.

PBXs may be cast directly into a casing or may be machined after curing to any desired geometry. The advantages of the charges include good mechanical properties, good charge safety and high thermal stability. Disadvantages include the use of toxic ingredients, long cure times, large numbers of ingredients in the system and, since the technology is relatively recent, new mixing and filling equipment may be required.

The Integrity of Military Explosive Fillings

Filling Difficulties

The vast majority of charges in current military use are melt-cast TNT-based fillings. A number of possible problems and defects may be encountered due to inadequate control of the filling process or inappropriate storage conditions. Some commonly encountered difficulties are described here.

Exudation

If an explosive contains an ingredient which becomes liquid in the temperature range normally experienced by a munition, it may in some circumstances be possible for the liquid to creep out of its casing via screw threads and other joints, and then contaminate the outside of the munition, giving rise to fire, toxic and possibly explosion hazards. This escape of the liquid from the explosive charge is called exudation. Such low melting point substances may have been intentionally incorporated in the explosive charge in the form of plasticisers or may simply be low melting impurities in a much larger quantity of a legitimate ingredient. Even non-explosives such as the plasticiser diethyl phthalate will have explosives dissolved in them and therefore would give rise to problems if exuded.

Low melting impurities in an explosive charge (for example, unsymmetrical TNT or dinitrotoluenes in a symmetrical TNT filling) will normally be distributed evenly throughout the charge and will be no trouble. Such impurities are generally more soluble in cold explosive than in warmer explosive and will tend to migrate down a temperature gradient. This phenomenon is used industrially to concentrate and then remove impurities from solids: it is called zone refining. In storage, munitions may be subjected to repeated temperature cycling. If a differential temperature can be set up along the munition, then impurities will be redistributed and concentrated at one end of the munition. This could be achieved if the store floor on which the munition stands keeps a fairly constant temperature but the air temperature changes by a greater extent. Its concentration may eventually become high enough for free liquid to form. If this liquid is at the end closed by a threaded plug, then the liquid can exude.

Another source of exudation is the incorporation of plasticisers in explosive charges. These may also be concentrated by zone refining or may literally be squeezed out of the charge if the charge is kept under pressure by anti-vibration pressure plates or other spring loaded retaining plates. Cognizance of this must be taken into account by warhead designers by incorporating 'O' ring seals to prevent the escape of plasticisers.

Other Charge Defects

Other defects which may arise as a result of inappropriate charge filling or handling and storage are illustrated in Figure 9.7. These could be discontinuities and cavities in the explosive charge which may serve as potential hotspots for premature initiation under the influence of the set back forces experienced by the charge when fired from a gun, or otherwise violently accelerated.

Large cavities may be formed if the explosive mixture is not completely melted and mixed before it is poured into the munition casing. Central cavities or annular ring cavities may occur due to piping as the TNT binder

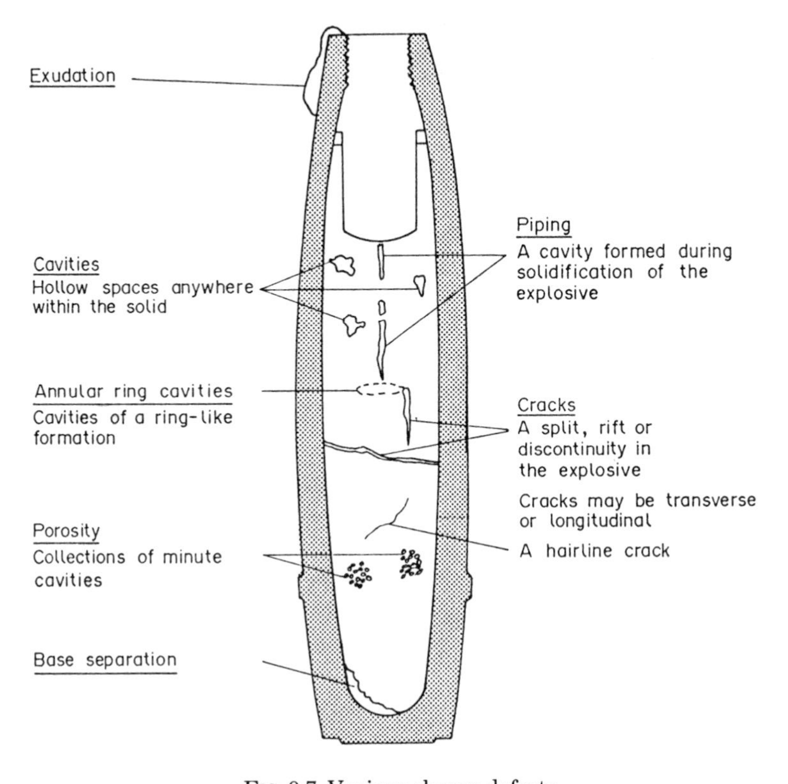

FIG. 9.7 Various charge defects

shrinks on setting, if care is not taken to control the cooling of the munition. The explosive melt needs to be degassed before being cast otherwise small gas bubbles will collect together as the charge cools and thereby create porosity.

Rough handling of munitions may result in cracks in the explosive charge. The frequency with which this occurs may be minimised by ensuring the maximum strength of the TNT binder is attained when the melt solidifies. The strength is controlled by the TNT crystal size and habit and this is determined by strict adherence to the optimum cooling regime and by the inclusion in the melt of substances which control the crystallisation of the TNT.

Bonding of the explosive charge to the munition case is an essential requirement for safety. Any separation between the charge and case will result in a cavity. To enhance the bond between the case for an explosive based on TNT, the inside of the case is coated with a bitumous paint. If this coating is imperfect or if moisture is present in the bottom of the case when it is being filled then adhesion will be poor in these regions and separation of the charge and case may result.

Finally, mention must be made of defects which arise as a result of the inclusion of foreign bodies in the melt. Grit, metal fragments including nuts, bolts, washers and splitpins have all been observed in radiographic inspection and result from inadequate control of the filling process.

Self Test Questions

Chapter 1

Q.1. Define an explosion.

Q.2. Name the three classes of explosions.

Q.3. What two essential requirements must an effective explosive meet?

Q.4. Which inventions greatly improved the safety and usefulness of explosives in the 19th century?

Q.5. In a typical black powder formulation, which ingredient is the oxidiser, the fuel and the sensitiser respectively?

Q.6. Calculate the oxygen balance of picrite ($CH_4N_4O_2$).

Q.7. Use the Kistiakowsky-Wilson rules (or the reversed rules as appropriate) to complete the equation for the explosion of PICRITE.

$$CH_4N_4O_2 \longrightarrow$$

Chapter 2

Q.1. A cube shaped grain of black powder 3 centimetres across each face was found to burn at a rate of 1 g per second under a given confinement. What would the rate of burning be under the same conditions if the 3 centimetre cube was subdivided into 1 centimetre cubes?

Q.2. Classify the following as secondary explosives, propellants or pyrotechnics: (a) TNT, (b) Dynamite, (c) Gunpowder, (d) ANFO, (e) Cordite, (f) Safety fuze, (g) Aero engine starter cartridge, (h) Illuminating flare.

Q.3. How much energy of a high explosive goes into the detonation shock wave?

Q.4. Name the factors which influence the detonation velocity of a secondary explosive.

Answers on page 165

Q.5. How fast does the 'jet' from a 'hollow charge' travel?

Q.6. What causes scabbing when an explosive is detonated against metal armour?

Chapter 3

Q.1. List the effects due to expansion of the hot gases from an explosion.

Q.2. Given the following information, list the explosives in descending order of their power.

Explosive	*Q (J/g)*	*V (l/g)*
TNT	3923	0.790
EGDN	6850	0.738
RDX	5691	0.856
PETN	5964	0.788

Q.3. What can be added to an explosive to increase its power?

Q.4. What are the various indices of explosive power or strength?

Chapter 4

Q.1. What is an explosive train?

Q.2. Give an initiator (primary) explosive which (a) always detonates and (b) is usually igniferous in use.

Q.3. Give two examples of explosives used as detonating boosters.

Q.4. What is the 'activation energy' as related to an explosive?

Q.5. List the ways in which initiation to burning may be achieved.

Q.6. How does percussion cause initiation of an explosive?

Q.7. List and describe the methods of igniferous electrical initiation.

Q.8. What explosives are at risk to accidental static discharge and how should they be handled?

Q.9. List the three methods by which explosives may be caused to detonate.

Answers on pages 165-167

Q.10. One of the answers to Question 9 is an electrical method, describe this method and give two advantages of using it.

Chapter 5

Q.1. What is the burning rate law or Vielle's equation?

Q.2. What is a typical value of the pressure index for a gun propellant?

Q.3. What is the name of the apparatus in which the burning rate of a propellant is measured?

Q.4. How does the grain size and shape of a propellant influence the burning rate?

Q.5. Some propellants display progressive burning. What is their geometric characteristic?

Q.6. What are the main ingredients of a triple-base propellant?

Q.7. What function does the ingredient, carbamite, perform in the composition of a double-base propellant?

Q.8. What are the main problems to be overcome in the design of caseless ammunition?

Q.9. What are the disadvantages of incorporating a flash inhibitor in a propellant composition?

Q.10. What is the lower explosion limit for carbon monoxide in propellant gases?

Chapter 6

Q.1. What are the different operating conditions for rocket propellants compared to gun propellants?

Q.2. What affects the value of specific impulse (I_s) for a rocket motor?

Q.3. Give three examples of liquid monopropellants.

Q.4. Give an example of a cryogenic (space-storable) and an earth-storable 'packaged' bipropellant.

Answers on pages 167-168

Q.5. List the advantages and disadvantages of liquid bipropellants.

Q.6. For solid rocket propellants, what is the difference between case-bonded and inhibited?

Q.7. What is an ABOL motor and why is it used?

Q.8. What are the advantages and disadvantages of double-base propellants for rocket motors?

Q.9. What is platonisation?

Q.10. How are double-base propellants modified and to what end?

Q.11. What are composite propellants?

Q.12. What are the advantages and disadvantages of composite propellants?

Q.13. What is a rubbery composite?

Q.14. What high explosive compound is added to composite propellants?

Chapter 7

Q.1. Define 'a pyrotechnic'.

Q.2. What are the four main categories of special effects which may be produced by pyrotechnics?

Q.3. What are the various roles of binders?

Q.4. Which atmospheric component causes most problems for the storage of pyrotechnics?

Q.5. In general, to what stimuli are pyrotechnic compositions most sensitive?

Q.6. List the factors which affect performance of compositions.

Q.7. What is a 'thermite' type composition?

Q.8. What is a first fire?

Answers on pages 169-170

Q.9. How are coloured smokes produced by pyrotechnic means?

Q.10. What composition gives the best illumination of long duration?

Q.11. How is coloured light produced?

Q.12. What is a decoy flare and what composition is used in these flares?

Q.13. What factors affect the mass extinction coefficient for a screening smoke?

Q.14. Give two types of screening smoke composition.

Q.15. How efficient are visible screening smokes at screening in the infra-red region (1-14)?

Chapter 8

Q.1. What type of devices are classed as pyromechanisms?

Q.2. What types of explosive filling are found in pyromechanical devices? Give examples of each type.

Q.3. How is the choice of high explosive filling influenced by the small size of pyromechanical devices?

Q.4. What methods can be employed to make detonating EED's safe from induced current caused by electromagnetic radiation such as radio waves, radar etc.?

Q.5. What is the difference between an actuator and a power cartridge?

Q.6. What explosive type normally is used in gas generators?

Q.7. Give some examples of high explosive containing pyromechanisms.

Chapter 9

Q.1. Give three reasons why TNT is a good binder for explosive formations.

Q.2. Name a common TNT processing system.

Q.3. What is the formulation of Torpex?

Answers on pages 170-171

Q.4. What is 'piping' in munitions filling?

Q.5. What is the optimum ratio of coarse, medium and fine particles in a melt-cast explosive filling?

Q.6. Name the three pressing techniques used in explosive charge production.

Q.7. Name three precured binder polymers used for pressed explosive charges.

Q.8. What does the term 'exudation' mean when applied to a round of ammunition?

Q.9. How is adequate bonding assured between a shell case and its explosive filling?

Q.10. What is the most common cause of separation of the charge from its case?

Answers on pages 170-171

Answers to Self Test Questions

Chapter 1

A.1. An explosion is the sudden release of high pressure gas.

A.2. Physical, Nuclear and Chemical.

A.3. It must have adequate SAFETY and RELIABILITY.

A.4. The developments of safety fuze, the fulminate detonator and dynamite.

A.5. Potassium nitrate is the oxidiser, charcoal is the fuel and sulphur is the sensitiser.

A.6. -30.76.

A.7. $CH_4N_4O_2 \longrightarrow CO + H_2O + H_2 + 2N_2$

Chapter 2

A.1. 3 g per second.

A.2. (a), (b), (d) secondary explosive, (e) propellant, (c), (f), (g), (h) pyrotechnic.

A.3. Approximately 50 per cent.

A.4. Density, confinement and heat explosion.

A.5. Up to 10,000 metres per second.

A.6. The reflection of the shock wave as a rarefaction at the rear metal/air interface.

Chapter 3

A.1. Lift, heave, cratering, propulsion of fragments, blast.

A.2. EGDN, RDX, PETN, TNT.

A.3. Aluminium.

A.4. Power index, Weight strength, Grade strength, Cartridge strength and Bulk strength.

Chapter 4

A.1. An explosive train is a series of explosive compositions joined together to give reliable initiation of the main charge. The first in the series is the initiator or primary composition, the second a booster and the third the main charge. These are in order of decreasing ease of initiation and increasing output.

A.2. (a) Lead or silver azide, (b) Lead styphnate or LDNR.

A.3. Tetryl, PETN or RDX.

A.4. Activation energy is the energy input required to cause the explosive to decompose thus releasing its energy.

A.5. External heat, flash or flame, percussion, friction, stabbing, electrical and chemical.

A.6. Percussion causes adiabatic heating of occluded air within the composition together with a certain amount of friction between explosive particles.

A.7.
1. Bridgewire or fuzehead – a resistance wire embedded in the explosive is heated by low tension electrical current raising the temperature to the ignition temperature of the explosive.
2. Conducting composition – the composition has added graphite or metal powder making it conduct electricity and having resistance becomes heated if a current is passed through it.

A.8. Primary explosives, nitrocellulose powders and some pyrotechnic compositions. When handled, the floors and benches should be covered by conducting material and workers should ensure they are in electrical contact with one of these surfaces. Furthermore, humidity should be above 65 per cent RH.

A.9.
1. Burning to detonation, as for lead azide.
2. Impinging of a detonation shockwave from a donor to an acceptor charge.
3. Exploding bridgewire (EBW).

A.10. The exploding bridgewire using a high tension pulse vaporises a resistance wire and the shockwave thus released directly detonates a secondary explosive. Two advantages are: (i) the device contains no primary explosive and is thus safer against accidental initiation, (ii) induced current could not initiate an EBW and thus radhaz problems are eliminated.

Chapter 5

A.1. $r = \beta p^{\alpha}$

A.2. 0.9 approximately.

A.3. The Strand Burner.

A.4. The grain size and shape determines the 'specific surface' of the grain. The burning rate is proportional to the specific surface.

A.5. Multitubular or perforated geometry.

A.6. Nitrocellulose, nitroglycerine and picrite.

A.7. It is a stabiliser.

A.8. The problems are how to obtain controlled fragmentation of the propellant to give a uniform burning rate and how to prevent cook-off.

A.9. Flash inhibitors tend to increase the propellant's smoke output.

A.10. 12.5 per cent.

Chapter 6

A.1. Rocket propellants burn at 15-30 MPa pressure compared to 450 MPa for gun propellants. Rocket propellants will generally have much longer burn times, seconds versus microseconds, and residues are much less of a problem, as is flame temperature.

A.2. $I_s = \frac{20n^{1/2}}{M}$ where Q is propellant energy (J g^{-1}), M

is the average molecular weight of the exhaust gases and n is the efficiency of the gas expansion process.

A.3. Hydrazine, nitromethane, iso-propylnitrate.

A.4. (a) Liquid hydrogen/liquid oxygen, (b) UDMH/IRFNA.

A.5. *Advantages:* High performance, clear exhaust, controllable thrust, often cheap.
Disadvantages: Complex plumbing, either cryogenic or toxic.

A.6. Case-bonded means that the propellant can be sealed directly to the motor case and thermal stresses do not detach the bond. Inhibited means that a flame-resistant sheath is put around the grain to prevent burning on that surface since that propellant will not case-bond.

A.7. ABOL stands for "all-burnt on launch' and is used for shoulder fired weapons to prevent injury to the firer from rocket efflux.

A.8. *Advantages:* smokeless exhaust, complex grain shapes are possible and they can be platonised.
Disadvantages: low temperature storage problems, extruded double-base cannot be case-bonded and cannot be aluminised to increase their modest performance.

A.9. Platonisation is the addition of, usually, a lead salt which gives a range of pressure where burning rate is independent of pressure.

A.10. Composite modified cast double-base propellants have aluminium and ammonium perchlorate added to increase specific impulse. Elastomer modified cast double-base propellants include an elastomeric polymer to the composition to improve physical properties at low temperature.

A.11. Composite propellants have a polymer fuel intimately mixed with a salt oxidiser, ammonium perchlorate. Aluminium is often included.

A.12. *Advantages:* good performance, case-bond well, can be aluminised to increase performance and generally have good physical properties.
Disadvantages: a very smokey exhaust and careful control of curing processes are required for the rubbery type.

A.13. A rubbery composite has as polymer fuel a system whereby the oxidiser can be added, the mixture put into a mould and then the polymer cross-links to give a thermoset polymer.

A.14. HMX (or possibly RDX).

Chapter 7

A.1. A pyrotechnic is a material capable of combustion when correctly initiated to provide a special effect. They are usually fuel/oxidiser mixtures.

A.2. Heat, light, smoke and sound.

A.3. Binders increase particle cohesion, coat and protect reactive ingredients such as metal powders, modify performance and modify sensitiveness.

A.4. Water is the main problem as it reacts with metal powders to produce hydrogen. Salts in the composition are often hygroscopic and attract the water vapour.

A.5. Friction, flame (flash) and spark.

A.6. Stoichiometry, burning rate additives, particle size and internal structure of the component particles.

A.7. A thermite composition contains a metal plus metal oxide which upon reaction produces heat as the oxygen swaps from one metal to the other. It relies on the difference between the heats of formation of the two metal oxides.

A.8. This is a composition pressed on to the main filling which picks up the flash of an igniter and then initiates the main filling.

A.9. A dye is vaporised by an intimately mixed cool burning composition, usually lactose/potassium chlorate, on reaching the atmosphere it recondenses to small particles giving a coloured smoke.

A.10. Magnesium/sodium nitrate/binder.

A.11. From excited ions; $SrCl^+$ (red), $BaCl^+$ (green).

A.12. A decoy flare is used to attract heat (infra-red) seeking missiles. The current optimum composition consists of magnesium/teflon/viton (MTV).

A.13. Mass extinction coefficient (α) depends upon transmission (τ) path length and concentration of smoke particles.

A.14. Phosphorus smoke and aluminium/hexachloroethane/zinc oxide.

A.15. Visible screening agents have poor infra-red screening properties, e.g. hexachloroethane smokes have $\alpha \sim 5\ m^2\ g^{-1}$ in the visible but only $\sim 0.1\ m^2\ g^{-1}$ in the infra-red.

Chapter 8

A.1. Although difficult to classify accurately, pyromechanisms include: mechanical actuators, gas generators and explosive separators.

A.2. (i) True pyrotechnics such as black powder or metal/salt oxidiser, (ii) explosive compounds of the primary explosive type such as LDNR, lead styphnate, KDNBF, etc., (iii) propellants, usually double-base or single-base, (iv) high explosives such as lead azide, PETN, RDX.

A.3. The high explosive compositions must have small critical diameters, thus limiting choice.

A.4. Use either a radio-frequency filtering device or use an exploding bridgewire detonator.

A.5. Terminology in this area is rather vague. An actuator will usually have the explosive content within the body of the device whereas a power cartridge will screw on to the mechanical part of the mechanism to power it.

A.6. Usually double-base propellant.

A.7. Explosive bolts, nuts and shackles, mild detonating fuze or cord.

Chapter 9

A.1. (a) It can be melted by steam heat. (b) It melts well below its ignition temperature. (c) It contributes to the overall explosives effect.

A.2. The Meissner process.

A.3. RDX, TNT, and aluminium.

A.4. It is the hollow ‘pipe’ left in an explosive charge due to the shrinking which occurs when a molten explosive solidifies.

A.5. 100:10:1.

A.6. Incremental, hydrostatic and isostatic pressing.

A.7. PTFE, KEL F and Viton rubber.

A.8. It refers to the escape of any low melting liquid impurity or plasticiser which may have been an ingredient of the explosive filling of the munition.

A.9. The inside of the shell is coated with a bituminous paint.

A.10. Presence of residual moisture in the munition case during the filling process often leads to poor adhesion and separation of the charge from its case.

Glossary of Terms

ABOL
All Burnt On Launch rocket motor.

Activation Energy
The minimum amount of energy which must be supplied to a chemical system to initiate a chemical reaction.

Adiabatic Change
A change undergone by a system such that no heat enters or leaves the system.

Adiabatic Flame Temperature (T_0 for a gun propellant)
The maximum temperature ideally achievable during combustion under adiabatic conditions.

AP
Ammonium perchlorate

Arrhenius Equation
An important equation in physical chemistry which expresses the way in which the rate constant of a chemical reaction is influenced by temperature.

Ballistic Modifier
Any additive incorporated in a propellant composition to alter its rate of burning.

Base Charge
The increment of secondary high explosive (commonly tetryl or PETN) in the base of a composite detonator, to enhance the shock wave produced by the primary explosive.

Binder
A wax, resin or polymer used to aid consolidation of a powdered explosive composition. Mostly used in pyrotechnics.

Bipropellant
A propellant system consisting of two liquids (fuel and oxidiser) which are mixed within the combustion chamber of the rocket motor.

Black Body Radiation
The energy spectrum emitted by a perfect absorber.

Booster

Any component of an explosive train which is interposed between the initiator and the main charge.

Bridgewire

An electrical filament which can be used to ignite a pyrotechnic or primary explosive which is in contact with it, normally inside an igniter or detonator.

Brisance

The shattering property exhibited by a detonating explosive.

Burning

The propagation of combustion by a surface process.

Candela (Cd)

The unit of luminous intensity.

Candle

A consolidated pyrotechnic composition.

Cap

A small metal container filled with a flame-producing explosive composition.

Case-bonding

The gas-tight adhesion of a rocket propellant grain to the walls of the combustion chamber of the motor.

CDB

Cast Double-Base propellant containing nitrocellulose and nitroglycerine.

Chalcogen

One of the elements; sulphur, selenium or tellurium.

CMCDB

Composite Modified Cast Double-Base. A rocket motor composition containing nitroglycerine, nitrocellulose and AP.

Combustion

An exothermic oxidation reaction producing flame, sparks or smoke. The oxidant may be part of the material as in a propellant, or oxygen from the atmosphere or other source.

Compatibility (as applied to explosives)

The possibility of one or more substances remaining in contact with an explosive without adverse physical or chemical effect on either.

Conducting Composition (CC)

An initiatory component (e.g. of a gun cartridge) in which the passage of an electric current through a small quantity of conducting explosive heats it to its ignition temperature. The use of a bridgewire is thus eliminated.

Cook-off

The initiation of an enclosed explosive by the conduction of heat through its container. Applicable to, for example, munitions exposed to conflagrations, or to live cartridges left loaded in hot guns.

Coolant

A substance added to propellants to reduce the flame temperature and hence the rate of burning.

Covolume

A correction applied to the equation of state to take account of the fact that at very high pressures (e.g. in a gun) the molecules of a gas occupy a finite volume. Each propellant has a slightly different covolume figure, measured in volume per unit mass.

CTPB

Carboxy-Terminated Polybutadiene, a polymer used in rubbery composite propellants.

Dautriche Test

A routine comparative method for measuring experimentally the velocity of detonation of explosives.

Decoy Flare

A flare fired to attract attention, usually of a heat-seeking missile.

Deflagration

A rapid burning in which convection often plays an important role.

Delay

A device incorporated into an initiation system so as to time the explosive event. Usually a pyrotechnic composition.

Detonating Fuze

Plastic tubing filled with powdered secondary high explosive, which will transmit a detonation wave over any required distance.

Detonation

An extremely fast explosive decomposition, in which an exothermic reaction wave follows and also maintains a shock front in the explosive.

Detonation Pressure

The dynamic pressure in the shock front of a detonation wave.

Detonator

An explosive device for starting detonation. It is small in size and may be designed for initiation by one of several methods, e.g. stabbing, bridgewire, etc.

Electro-Explosive Device (EED)

Any explosive device initiated by electrical input.

EMCDB

Elastomer Modified Case Double-Base, a rocket propellant composition containing a polymer to modify the physical properties of a CDB.

Eutectic

A mixture of two or more materials in the proportions which give the lowest melting point.

Exothermic

Giving out heat energy.

Exploding Bridgewire (EBW)

A detonator in which an electric filament is explosively vaporised by a high tension pulse, and thereby sets up detonation in a surrounding secondary explosive filling. The use of primary explosive is thus eliminated.

Explosion

A violent expansion of gas.

Explosiveness

The rate at which a particular explosive, when exposed to a given stimulus, gives up its energy, and/or the degree to which it does so.

Explosive Power

The work capacity of an explosive, usually referring to high explosives. It is not the *rate* of doing work, although in practice the rate may affect the experimental measurement of values. Explosive power can be calculated as a percentage of the work done by a standard explosive (commonly picric acid or blasting gelatine) on the basis of the amount of heat and gas generated. Experimental measurement by the lead block test or ballistic mortar generally correlates well with calculation.

Explosive Train

An arrangement used to lead explosive reactions from one place to another. The distance may be minimal, e.g. the assembly of fuze detonator, magazine, exploder and main charge in a shell or bomb. Alternatively it may be spread over a distance, e.g. by using detonating fuze in a demolition. Regardless of its dimensions, an explosive train may be one of two kinds - igniferous or disruptive.

First Fire
A pyrotechnic composition used to ignite a main charge.

Flare
A pyrotechnically generated source of light (and heat).

Flash
A short-lived flame used as a method of initiation.

Flash-receptive Detonator
A type of detonator which is initiated by a flash.

Force, or Force Constant
The work capacity of an explosive, usually applied to gun propellants. It can be calculated for high explosives or propellants and in the case of the latter it can be determined experimentally by the use of the Closed Vessel.

Form Function
The mathematical function describing how the surface area of a propellant grain of particular shape changes in the course of combustion.

Fragmentation
The shattering effect of an explosive upon its container, e.g. fragmentation of a shell case by its explosive filling.

Friction Test
A test to assess the susceptibility of the explosive to initiation by friction.

Fuze
(a) Cord or tube for the transmission of detonation or flame.
(b) A compact, engineered assembly of explosive components, including safe-arm devices, as a means of initiating a munition.

Note:
The spelling FUZE has been standardised in British service literature for items under both definitions. Most other literature uses FUSE for items under a and b.

Grain (as applied to propellants)
A discrete piece of a propellant composition, usually of a precise geometrical shape, and irrespective of size.

HARM
High-speed anti-radiation missile.

Heat Capacity
The amount of heat absorbed by a substance when its temperature is raised by 1K. The unit of substance may be one of mass, giving a 'specific heat capacity'.

Heat of Explosion
The amount of energy released when one mole or unit mass of explosive burns or detonates under adiabatic conditions.

High Explosive
An explosive which is capable of detonation under the normal conditions of use.

HMX
A nitramine secondary high explosive similar to RDX but with a higher performance. Also called Octogen or tetramethylene tetranitramine.

HNS
Hexanitrostilbene. A secondary high explosive used in special applications.

Hollow Charge
A charge, usually cylindrical, having a cavity opposite the point of initiation to exploit the Munroe effect. The cavity may be lined with metal to enhance the effect.

Hot Spot
A small, localised region in an explosive substance which is characterised by a temperature much higher than that of its surroundings. This is relevant to mechanisms of initiation.

HTPB
Hydroxy-Terminated Polybutadiene, a polymer used in rubbery composite compositions.

Hygroscopicity
The property of a substance of absorbing and retaining water from the atmosphere.

Hypergolic
Term describing certain liquid bipropellant combinations which ignite spontaneously on mixing.

Igniter
A device for starting the burning process in an explosive train.

Impact Test
A test to assess the susceptibility of the explosive to initiation by impact.

Incendiary
A pyrotechnic composition designed to give a large steady heat output to cause fire or damage structural materials.

Initiator

A device for igniting or detonating explosives or pyrotechnics, commonly an igniter or detonator.

Insensitiveness, Figure of (F of I)

A figure relating to a particular explosive which indicates its comparative insensitiveness to mechanical impact as determined in the Rotter Test. The higher the figure, the *less* sensitive is the explosive to this form of initiation.

Intermediary

An explosive of which the sensitiveness to initiation by impact or shock wave is intermediate between that of primary explosives and secondary explosives.

Kistiakowsky - Wilson Rules (K-W Rules)

A set of empirical rules for calculating approximately the composition of the gaseous products of a high explosive.

Lead Block Test (or Trauzl Test)

See Explosive Power.

Lumen

The luminous flux unit. It is equal to the luminous energy radiated per second per steradian by a uniform source of one candela luminous intensity.

Luminous Output

As for the definitions of radiant output but now for visible light.

Molar (or Universal) Gas Constant (R)

A constant which appears in the equation of state for a perfect gas, viz. $PV=RT$. Its value is the same for all gases.

Monopropellant

A liquid that burns in the absence of air to produce hot gas for propulsion purposes.

Munroe Effect (or Neumann Effect)

A local concentration of shock wave energy which occurs when the wave emerges from a detonating charge via a re-entrant shape in the charge surface.

Oxygen Balance (Ω)

For an explosive containing the usual elements carbon, hydrogen, nitrogen and oxygen, it is the percentage by weight of oxygen, positive or negative, remaining after combustion, assuming that all the carbon and hydrogen is converted into carbon dioxide and water.

PBAN

Polybutadiene-Acrylonitrile, a co-polymer used in composite propellants.

PETN

Pentaerythritol Tetranitrate, a secondary high explosive used as the filling for detonating cords and as a booster in detonating explosive trains.

Photoflash

A pyrotechnic store designed to produce millions of candela in the millisecond time scale.

Plasticiser

A substance added to double-base propellants to assist the gelatinisation of the nitrocellulose content and reduce the chances of grain fragmentation when burning.

Platonisation

The modification of the burning rate versus pressure curve of a double-base rocket propellant to achieve constant burning rate over a given range of pressure.

Pressure of Combustion (or of Explosion)

The maximum static pressure produced when a given weight of high explosive or propellant is burned, without detonating, in a closed vessel of given volume.

Primary Explosive

An explosive which is readily ignited or detonated by a small mechanical or electrical stimulus.

Propellant

An explosive used to propel a projectile or missile, or to do other work by the expansion of high pressure gas produced by burning, e.g. for starting engines.

PTFE

Polytetrafluoroethylene, a fluorinated organic polymer, $(-(CF_2)_{n}-)$.

PU

Polyurethane, polymer used in rubbery composite compositions.

Pyrotechnic

A material capable of combustion when correctly initiated to provide a special effect. They are usually mixtures of fuel and oxidiser.

Quickness

A measure of both the pressure generated in a confined space by a propellant of given composition and grain shape, and the speed with which the pressure is produced. Effectively the product of force constant and vivacity.

Radhaz (Radio Frequency Hazard)
The danger of the accidental initiation of an electro-explosive device by radio-frequency electromagnetic radiation.

Radiant Emittance
Radiant flux per given area of emitter (source).

Radiant Energy
Total energy emitted by an electromagnetic source (light or heat).

Radiant Flux
Rate of radiant energy production.

Radiant Intensity
Radiant flux per given solid angle from a point source.

Rate of Burning
(a) The rate of regression of the burning surface of an explosive, usually a propellant grain, in length per unit time, under given conditions of pressure and grain temperature. Designated r and sometimes referred to as the 'linear rate of burning'.
(b) The rate of consumption of a burning explosive, usually a propellant or bipropellant combination, in terms of mass per unit time. Designated m.

RDX
A secondary high explosive also called Cyclonite or trimethylenetrinitramine. Much used in high explosive formulations, for example: RDX/TNT or RDX/wax.

Rotter Test
An empirical test in the United Kingdom to determine the sensitiveness of a given explosive to mechanical impact, in this case a falling weight. An important part of hazard assessment. Results are calculated as a *Figure of Insensitiveness*. In the United States the Fallhammer Test is used.

Secondary Explosive
An explosive which can be made to detonate when initiated by a detonation wave or other shock front but which does not normally detonate when heated or ignited.

Sensitiveness
A measure of the relative ease with which an explosive may be ignited or initiated by a particular stimulus.

Set-back
The inertial forces acting upon the contents of a projectile due to its acceleration in a gun barrel on firing.

Shaped Charge

This term usually implies some application of the Munroe effect in the geometry of a HE charge.

Shock Front

A discontinuous change in the pressure and other parameters of a medium. This change is propagated at supersonic speed.

Shock Wave

A shock front, together with its associated phenomena. It can be transmitted through all kinds of matter and can be generated by explosive or other means.

Simulator

A device used in training to simulate battlefield light and noise.

Smoke

An aerosol of particles which settle very slowly in air.

Spalling

The production of a fragment or fragments from the rear face of a solid material which has been subjected to a detonation shockwave on the front face.

Spark Test

A test to assess the susceptibility of an explosive to initiation by electric discharge.

Specific Impulse

A measure of the output of a rocket motor.

Stabiliser

A substance which prevents or reduces autocatalytic decomposition of explosives.

Stand-off

The distance of an explosive charge from a target at the instant of detonation. There may be an optimum value at which best performance is achieved.

Steradian

Unit solid angle; there are 4π steradians in a sphere.

Stoichiometric Mix

A mixture of chemical reactants in which the quantities of each component are such that they are balanced and all material reacts.

Stoichiometry
The relative quantities of components in a reacting system.

Strand Burner
A device for measuring experimentally the rate of regression of the burning surface of a length of propellant, burning endwise at a fixed pressure.

Surface Moderant
An additive applied to the surface of some gun propellants to reduce the burning rate in the early stage of combustion.

Temperature of Ignition
The temperature at which an explosive ignites under specified conditions, e.g. the rate of heating.

Tetryl
A secondary high explosive used as a booster in detonating explosive trains.

Thermal Explosion
Explosion resulting from exothermic reaction in an explosive charge in a region where heat is liberated more rapidly than it can be transferred.

Thermate
A generic term for pyrotechnic compositions used to fill incendiary bombs.

Thermite Composition
A mixture of metal powder and a metal oxide providing heat in a gas-less reaction.

Thermitic Reaction
Usually the reaction between a metal powder and metal oxide in a pyrotechnic composition.

TNC
Tetranitrocarbazole, an organic compound used as energetic fuel in some pyrotechnic compositions.

TNT
Trinitrotoluene, an insensitive secondary high explosive used in combination with RDX.

Tracer Composition
A pyrotechnic mixture designed to withstand the spin and acceleration of shells or bullets. It is used to give illumination so that the firer can monitor the flight of the projectile.

Tracking Flare
A relatively long burning pyrotechnic filling used to allow visual or instrumental tracking of missiles.

Velocity of Detonation

This is the speed at which a detonation wave progresses through an explosive. When, in a given system, it attains such a value that it will continue without change, it is called the Stable Velocity of Detonation for that system.

Vivacity

The way in which a gun propellant charge of given composition and grain dimensions behaves in respect of its mass burning rate during combustion under adiabatic conditions. Experimental determination is necessary and employs the Closed Vessel.

Index